W9-AGM-892

St. Clair College

JAN 8 1982

Science Library

An Introduction to Synthesis Using Organocopper Reagents

An Introduction to Synthesis Using Organocopper Reagents

GARY H. POSNER

Department of Chemistry
The Johns Hopkins University

A WILEY-INTERSCIENCE PUBLICATION

JOHN WILEY & SONS
New York • Chichester • Brisbane • Toronto

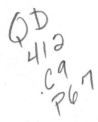

QD
412
.C9
P67

Copyright © 1980 by John Wiley & Sons, Inc.

All rights reserved. Published simultaneously in Canada.

Reproduction or translation of any part of this work
beyond that permitted by Sections 107 or 108 of the
1976 United States Copyright Act without the permission
of the copyright owner is unlawful. Requests for
permission or further information should be addressed to
the Permissions Department, John Wiley & Sons, Inc.

Library of Congress Cataloging in Publication Data:

Posner, Gary H.
  An introduction to synthesis using organocopper
reagents.

  "A Wiley-Interscience publication."
  Includes index.
  1. Organocopper compounds. 2. Chemistry,
Organic–Synthesis. I. Title.

QD412. C9P67   547′.2   80-13538
ISBN 0-471-69538-6

Printed in the United States of America

10 9 8 7 6 5 4 3 2 1

To
Rose and Joe,
who helped make it all possible

Rhoda, Joey, and Michael,
who helped make it all worthwhile

E. J.,
who sees and solves the big problems
with perseverance, innovation, and style

The synthesis of substances occurring in Nature, perhaps in greater measure than activities in any other area of organic chemistry, provides a measure of the condition and power of the science. For synthetic objectives are seldom if ever taken by chance, nor will the most painstaking, or inspired, purely observational activities suffice. Synthesis must always be carried out by plan, and the synthetic frontier can be defined only in terms of the degree to which realistic planning is possible, utilizing all of the intellectual and physical tools available. It can scarcely be gainsaid that the successful outcome of a synthesis of more than thirty stages provides a test of unparalleled rigor of the predictive capacity of the science, and of the degree of its understanding of its portion of the environment. Since organic chemistry has produced syntheses of this magnitude, we can, by this yardstick, pronounce its condition good.

R. B. Woodward in A. R. Todd (Ed.), *Perspectives in Organic Chemistry,* Interscience, New York, 1956, p. 155.

The synthetic chemist is more than a logician and strategist; he is an explorer strongly influenced to speculate, to imagine, and even to create. These added elements provide the touch of artistry which can hardly be included in a cataloguing of the basic principles of Synthesis, but they are very real and extremely important. Further, it must be emphasized that intellectual processes such as the recognition and use of synthons require considerable ability and knowledge; here, too, genius and originality find ample opportunity for expression.

The proposition can be advanced that many of the most distinguished synthetic studies have entailed a balance between two different research philosophies, one embodying the ideal of a deductive analysis based on known methodology and current theory, and the other emphasizing innovation and even speculation. The appeal of a problem in synthesis and its attractiveness can be expected to reach a level out of all proportion to practical considerations whenever it presents a clear challenge to the creativity, originality and imagination of the expert in synthesis.

E. J. Corey in *Pure Appl. Chem.,* 1967, **14**, 30.

I would not like to neglect to mention that we are still very poor in synthetic methods in organic chemistry. Most of the synthetic work is done with organic reactions of the type which have been known for a long time. If you know 20 organic reactions you probably know most of the steps used in synthetic work, particularly in industry, but I am quite sure there must be hundreds of other organic reactions to be discovered. We have not in the past thought about these problems in the right way. When we have been faced with a problem of effecting a chemical synthesis we have sought known methods. We have not paused to think why we do not invent a new method every time. If we adopt this philosophy we are going to be extremely busy till the end of the century (a) trying to equal the enzymes, and (b) thinking up new ways of synthesis.

D. H. R. Barton in *Chem. Britain*, 1973, **9**, 149.

Foreword

The microcosm of organic chemistry has for over a century revealed to the explorer new ideas, unsuspected objects, and worlds unforeseen. After a first reading of this exposition of organocopper chemistry and its applications by Professor Gary Posner, because of the contrast between the transformations of matter dealt with in this book and the chemistry I had learned and practiced as a student in the late 1940s, I was reminded of the curious simile between chemistry and astronomy. The origins of the new chemistry of organocopper reagents are of considerable interest as much because of their unusual nature as for their far-reaching consequences.

Henry Gilman, that dauntless pioneer of so much important organometallic chemistry, prepared the first organocopper reagents [H. Gilman and J. M. Straley, *Rec. Trav. Chem.*, **55**, 821 (1936)], for which reason we have favored calling the whole class "Gilman reagents." His was an exercise in the purest form of fundamental research, compelled mainly by curiosity. Another formidable figure, Morris S. Kharasch, during his systematic investigations of the effect of transition metals on the reactions of Grignard reagents, discovered the famous and highly useful catalysis by cuprous ion of the conjugate addition of organomagnesium compounds to enones [M. S. Kharasch and P. O. Tawney, *J. Am. Chem. Soc.*, **63**, 2308 (1941)]. In the mid-1960s H. O. House and colleagues [H. O. House, W. L. Respess, and G. M. Whitesides, *J. Org. Chem.*, **31**, 3128 (1966)] provided unambiguous experimental evidence that Gilman reagents were involved in the Kharasch conjugate addition process. At about the same time our group at Harvard was much involved in coupling and cross-coupling reactions of organonickel reagents as a synthetic method. As a consequence of the discovery of specific limitations in the use of nickel reagents, and of the knowledge of the classical copper-catalyzed Ullman reaction, attention was turned to the study of cross-coupling reactions of electrophiles such as carbon halides with Gilman reagents. This led to the key discoveries of Dr. Posner [E. J. Corey and G. H. Posner, *J. Am. Chem. Soc.*, **89**, 3911 (1967); **90**, 5615 (1968); G. H. Posner, Ph.D. thesis, Harvard University, 1968]. Further advances came at a breathtaking pace, with many outstanding contributions which are detailed on the pages that follow.

Gary Posner has been one of the principal figures in the field ever since his graduate days. Not surprisingly, he has provided us with an authoritative but eclectic and critical summary of syntheses using organocopper chemistry. This book is bound to be of great value to beginning and advanced students alike. It also draws attention to another aspect of synthesis which has assumed major significance in pedagogy and practice in the past 15 years, the logic of synthesis and especially the retrosynthetic analysis of synthetic problems.

It is safe to predict that this small volume will provide much impetus for the further development of organometallic syntheses and lead as well to the unveiling of some new surprises in the chemical microcosm.

E. J. COREY

Harvard University
Cambridge, Massachusetts
January 1980

Preface

The goal of this book is to describe organocopper chemistry in a fast, easily readable, and introductory fashion. Organized like my two *Organic Reactions* reviews on this subject, organocopper addition and substitution reactions are discussed here in terms of a series of natural product syntheses achieved since 1970. This format has been chosen carefully to impress you, the reader, with the usefulness of organocopper reagents in the synthesis of many different types of substances beneficial to mankind: food flavors and dyes, anti-inflammatory and antibiotic steroid drugs, perfume fragrances, prostaglandins, an anticancer compound, a sugar, hashish, vitamin E, insect growth-regulators, and insect sex-attractant pheromones. These recent examples make a strong case for the vitality and fundamental importance specifically of organocopper chemistry and generally of synthetic organic chemistry in our modern society. Inspired by students' enthusiastic responses to practical applications given in my introductory organic chemistry and organic synthesis courses at Johns Hopkins, I have included here brief discussions of the relevance of organic synthesis to the food, dye, perfume, pharmaceutical, and agricultural industries.

The second goal of this book is to describe the strategy of organic synthesis in a clear, introductory fashion. Emphasis is on carbon-carbon bond formation exclusively rather than on the interchange of functional groups. Retrosynthetic analysis is used to illustrate logical consideration of a synthetic problem, and you are encouraged at various places in the text to put this book down and to work with pencil and paper designing your own rational approaches to the construction of target molecules. It is hoped that you will be stimulated in this way to compare your ideas with those recorded in the primary literature, for which references are given at the end of each chapter, and to increase your skill in designing organic syntheses.

This book is intended as general reading or as part of a course in organometallic or synthetic chemistry for organic chemistry graduate and advanced undergraduate students. Industrial and academic chemists may read this monograph to acquaint themselves with the strategy of organic synthesis or to gain a broad perspective (and possibly some research ideas) in organocopper chemistry before studying review articles or the primary literature for full experimental details and information on the scope and limitations of organocopper chemistry.

GARY H. POSNER

Baltimore, Maryland
August 1980

Acknowledgments

I thank my friend Mark S. Fischer for being the first to suggest and encourage my writing a monograph on organocopper chemistry; he planted the original seed which has now germinated into this book. Clive Henrick and John Baum of the Zoëcon Corporation provided valuable information on insect pheromone synthesis, and William Galetto and William Zeiger of the McCormick Spice Company provided valuable information on food flavors. Rose K. Rose and Kevin A. Babiak read the manuscript and provided some valuable corrections. I thank Theodore Hoffman of John Wiley & Sons for his patience in waiting for me to complete writing this book, and I thank Barbara MacConnell-Hamilton for her patience and skill in typing the manuscript.

G. H. P.

Contents

An Introduction to Synthesis Using Organocopper Reagents

1

Introduction

Organic synthesis generally involves forming or breaking a bond between two carbon atoms and changing one functional group into another. Formation of a carbon-carbon σ-bond using an organometallic reagent has been performed routinely for many years with metals such as magnesium (Grignard reagents), lithium, sodium, and zinc. Transition metals, however, such as palladium, titanium, chromium, iron, cobalt, nickel, and copper are now being used more and more frequently to connect two carbon units efficiently and under especially mild reaction conditions (e.g., at or below room temperature). Organocopper reagents in particular have become popular among organic chemists for use in synthesis of larger molecules from smaller ones. This widespread acceptance and use of organocopper reagents are largely attributable to their ease of preparation and to their ability to effect transformations difficult or impossible to accomplish effectively with any other reagents. Organocopper reagents often react in a highly selective fashion: stereoselectively, regioselectively, and chemoselectively. These characteristics are especially important in the construction of complex organic molecules, as illustrated by the syntheses discussed later in this book. Several detailed reviews of organocopper chemistry are listed at the end of this chapter.

A. A Brief History of Organocopper Chemistry

Before the mid-1960s organocopper chemistry was studied only in sporadic fashion. Gilman and his students prepared methylcopper in 1936. In 1941 it was observed that the normal course of reaction

1

between a Grignard reagent and a cyclohexenone (i.e., 1,2-addition to the carbonyl group) was changed to 1,4-addition by the presence of a catalytic amount of copper(I). In 1952 Gilman and his students prepared dimethylcopperlithium (lithium dimethylcuprate). Not until the mid-1960s, however, did intensive research in the United States (at M.I.T. and Harvard) and in France reveal the broad synthetic potential of organocopper reagents.

In most cases organocopper reagents are prepared by adding an organomagnesium or an organolithium reagent via syringe to a copper(I) species at low temperature in an inert solvent and under an inert atmosphere. Less frequently, organocopper reagents are prepared by starting with organozinc, organoboron, or organozirconium compounds. Very recently, some organocopper compounds have been prepared from organic halides and highly reactive copper powder.

There are several different types of useful organocopper reagents, as shown in Table 1-1. The chemical behavior of one type of organocopper compound is sometimes very different from that of another type, and the same organocopper reagent often reacts differently in different solvents. It is often necessary, therefore, to examine empirically several types of organocopper species and to use different solvents and temperatures to arrive at the optimum choice of reagent and conditions for a desired transformation. Nevertheless, based on experience during the past 15 years, some sound generalizations can be made, as follows: (1) catalytic organocopper compounds are most effective when prepared from Grignard rather than organolithium reagents; (2) pure stoichiometric organocopper compounds (RCu) are less reactive than the corresponding cuprate species (R$_2$CuLi), and among pure RCu species only cuprous acetylides are particularly useful in substitution reactions; (3) phosphorus and sulfur ligands often solubilize organocopper compounds and sometimes allow very pure organocopper(I) compounds to be prepared, but these ligands sometimes make product isolation difficult; (4) homocuprates and mixed homocuprates are by far the most popular types of organocopper reagents; (5) organo(hetero)-cuprates are used for special reactions in which the organic group is particularly valuable, or in which an unusually stable organocopper reagent is needed; (6) R$_3$CuLi$_2$ and related complex organocopper-lithium species are relatively new and have been used thus far mainly for addition reactions; and (7) organocopper-borane complexes, devel-

Table 1-1 Organocopper Reagents

Types of Organocopper Reagents	General Name
RMet + CuX(catalytic)	Catalytic organocopper compound
RCu	Stoichiometric organocopper compound
RCu · Ligand	[Ligand = R_3P, $(RO)_3P$, R_2S]
R_2CuMet (Met = Li, MgX)	Homocuprate
R_2CuMet · Ligand	
RR'CuMet	Mixed homocuprate
R(Z)CuMet (Z = OR', SR', CN, Cl, Br)	Organo(hetero)cuprate or heterocuprate
R_nCuLi$_{n-1}$ ($n > 2$)	Complex organocopperlithium compounds
RCu · BF_3⎫ RCu · BR_3⎭	Organocopper-borane complexes

$$RMet + CuX \longrightarrow RCu + MetX \text{ or } R(X)CuMet$$
$$RMet + CuX \cdot Ligand \longrightarrow RCu \cdot Ligand + MetX$$
$$RCu + RMet \longrightarrow R_2CuMet$$
$$RCu + R'Met \longrightarrow RR'CuMet$$
$$RMet + CuZ \longrightarrow R(Z)CuMet$$

Scheme 1-1 Preparation of Organocopper Reagents

oped recently in Japan, are particularly useful for conjugate addition to acetylenic carbonyl compounds and for substitution of allylic electrophiles (reactively: $R_2CuLi > RCu > RCu \cdot BR_3$). Many specific examples supporting and extending these generalizations are given in syntheses 1–33 on the following pages.

Their thermal instability (see synthesis 33) and their high reactivity toward oxygen and water make isolation and structural characterization of most organocopper species very difficult. Lithium dimethylcopper, as representative of the homocuprates, has been shown by molecular weight measurement (vapor pressure depression), by NMR spectroscopy, and by solution X-ray scattering methods to exist in diethyl ether solution as a dimer represented in the accompanying structure. Likewise, lithium diarylcopper species in which the aryl group carries a coordinating heteroatom side arm have been shown by Dutch researchers using NMR and X-ray methods to exist as Cu_2Li_2 clusters with aryl groups bridging copper and lithium atoms via 3-center-2-electron bonds. Such aggregates are not true "ate" complexes of positive and negative ions; although the name "cuprate" therefore conveys inaccurate structural information, "lithium cuprate" is commonly used and is used here interchangeably with "diorganocopperlithium" and "lithium diorganocopper."

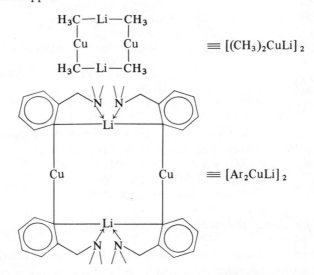

$$H_3C—Li—CH_3$$

$$\equiv [(CH_3)_2CuLi]_2$$

$$\equiv [Ar_2CuLi]_2$$

The structures of some fluorinated stoichiometric organocopper compounds (ArCu) were determined at DuPont by kinetic and molecular weight (e.g., mass spectrometry) methods to be tetrameric and in some cases octameric aggregates.

B. Typical Reactions of Organocopper Reagents

The two most popular reactions of organocopper reagents are (1) additions to carbon-carbon double and triple bonds, and (2) substitutions of organic halides and alcohol derivatives. Thermal and oxidative dimerization of organocopper compounds is of some use, especially for conversion of copper acetylides into conjugated diynes (Glaser reaction) and for formation of symmetrical biaryls.

Detailed reviews of organocopper additions to carbon-carbon multiple bonds are listed at the end of this chapter. Although there are a few examples of organocopper addition to olefins and acetylenes activated by nitro, phosphonyl, and sulfonyl groups, in this book we focus attention on the more common cases of organocopper addition to olefins and acetylenes activated by carbonyl groups. We also comment on organocopper addition to unactivated acetylenes (synthesis 10). α,β-Unsaturated carbonyl compounds react with most types of organocopper reagents in a conjugate manner, leading, after aqueous work-up, to a larger molecule in which the new carbon-carbon bond has been formed β to the carbonyl group. Organocopper conjugate addition is therefore like the Michael reaction of enolate ions with unsaturated carbonyl compounds in terms of the *position* of the new carbon-carbon bond and in terms of sensitivity to *steric factors* in the carbonyl compound, but it is unlike the Michael reaction in that a hydrocarbon group becomes attached at the β-carbon, producing a new *mono*carbonyl compound rather than 1,5-*di*carbonyl compound. In contrast to organocopper reagents, Grignard and organolithium compounds usually react with α,β-unsaturated carbonyl compounds via 1,2-addition to the carbonyl group. Organocopper reagents are generally less basic than the corresponding organolithium and organomagnesium reagents; enolization of carbonyl compounds by organocopper reagents is rare. Unsaturated *ketones* are more reactive than the corresponding unsaturated *esters* toward organocopper reagents, and *acetylenic* carbonyl compounds are more reactive than the corresonding *ethylenic* carbonyl compounds. Other important aspects of organocopper addition reactions are illustrated and discussed in syntheses 1-11.

Reviews of organocopper substitution reactions are listed at the end of this chapter. A generalized reactivity series of electrophiles toward lithium diorganocopper reagents is as follows: acid chlorides > aldehydes > tosylates $\approx$ epoxides > iodides > bromides > chlorides > ketones > esters > nitriles >> alkenes. The relative reactivity of alkyl halides is allylic > primary > secondary >> tertiary, and the kinetics of displacement are roughly first order in organocopper reagent and first order in substrate. Based on these relative reactivity series, chemoselective reaction of organocopper reagents has been achieved in many cases with one functionality in a bifunctional or polyfunctional compound. Organocopper substitution with alkyl halides is like the Wurtz coupling reaction (RX + Na) in that two *hydrocarbon* groups are joined, but it is unlike the Wurtz coupling in that union of two *different* hydrocarbon groups R (from R_2CuLi) and R' (from R'X) can be achieved without concomitant formation of the undesired symmetrical dimer R'R'.

The organocopper substitution reaction can be accomplished with a wide variety of nucleophilic R groups: R = alkyl (primary, secondary, tertiary), alkenyl, aryl, heteroaryl, alkynyl, allylic, and with many R groups carrying remote functionalities (e.g., ethers, acetals, ketals, sulfides). The stereochemical result of this substitution reaction is inversion of configuration with secondary alkyl halides and retention of configuration with vinylic halides. In all known examples the nucleophilic R group is transferred from the organocopper reagent with retention of configuration. In contrast to organocopper reagents other highly nucleophilic organometallic reagents (e.g., lithium and magnesium alkyls) are usually also strongly basic and therefore often undergo other reactions besides substitution. Other important aspects of organocopper substitution reactions are discussed in syntheses 12-33.

In multifunctional molecules that are both alkyl electrophiles and α,β-ethylenic carbonyl compounds, the relative amounts of organocopper substitution and conjugate addition can be changed by changing the alkyl-X leaving group and by changing the solvent. For example, we have shown that the accompanying enone mesylate, X = Ms, undergoes mesylate displacement by dimethylcopperlithium, whereas the corresponding enone bromide, X = Br, undergoes mainly conjugate addition (followed by intramolecular displacement). House and his co-workers recommend diethyl ether as solvent for organocopper conjugate addition reactions and diethyl ether-hexamethylphosphoramide (purified by distillation from BaO) for organocopper substitutions with alkyl halides. Most organocopper substitutions proceed well with tetrahydrofuran (THF) as solvent. Because there are some exceptions to these generalizations, however, review articles or the primary literature should be consulted before choosing a solvent for a particular organocopper reaction.

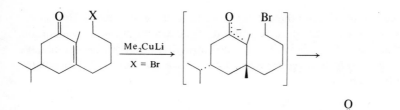

C. Mechanisms

Despite substantial efforts since 1965, the *detailed* mechanisms of organocopper addition and substitution reactions remain uncertain. There is widespread agreement, however, about the *general* mechanism of organocopper substitution reactions. Oxidative *trans*-addition of a d^{10} cuprate to an R'X electrophile produces a planar copper(III) (d^8) intermediate whose lifetime is influenced by the nature of X; reductive *cis*-elimination leads to the unsymmetrical coupling product RR' and the RCu(I) and LiX as shown in the following equation:

$$
\overset{I}{R_2CuLi} \xrightarrow{R'X} \left[R-\overset{\overset{\displaystyle R'}{|}}{\underset{\underset{\displaystyle X}{|}}{Cu^{III}}}-R \right] Li \longrightarrow RR' + RCu(I) + LiX
$$

Although there is widespread agreement about the likelihood that transient copper(III) intermediates are involved also in organocopper conjugate addition reactions, there are differences of opinion regarding how the copper(III) intermediates are formed. Direct nucleophilic oxidative addition of an organocopper(I) compound to the β-carbon atom of an α,β-ethylenic carbonyl compound produces a copper(III) intermediate; indirect oxidative addition proceeds via electron transfer from the organocopper(I) compound to the carbonyl substrate, generating a copper radical cation and an enone radical anion which then combine to produce the copper(III) intermediate. Reductive elimination causes attachment of an R group to the β-carbon atom and generation of an enolate, as shown in the accompanying equation. Unfortunately, there is only indirect evidence thus far for copper(III) intermediates in these reactions.

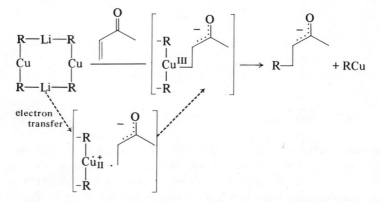

   As a more exact picture of the mechanisms of organocopper sub-stitution and addition reactions develops, there will emerge a more thorough understanding of why copper is one of the most effective transition metals in promoting carbon-carbon σ-bond formation. Undoubtedly, the uniqueness of copper is attributable in large part to the relatively low ionic character of a copper-carbon bond, to the low oxidation potential (0.15 V) separating cuprous from cupric ions, and to the tendency of copper to form polynuclear copper clusters and mixed-valence copper compounds.

Organocopper Reviews

Posner, G.H., "Conjugate Addition Reactions of Organocopper Reagents," *Org. React.,* 1972, **19**, 1.

Posner, G. H., "Substitution Reactions Using Organocopper Reagents," *Org. React.,* 1975, **22**, 253.

Normant, J. F., "Organocopper(I) Compounds and Organocuprates in Synthesis," *Synthesis,* 1972, 63.

Normant, J. F., "Organocopper Reagents in Organic Synthesis," *J. Organo-met. Chem. Lib.,* 1976, **1**, 219.

Normant, J. F., "Stoichiometric *vs.* Catalytic Use of Copper(I) Salts in the Synthetic Use of Main Group Organometallics," *Pure Appl. Chem.,* 1978, **50,** 709.

Jukes, A. E., "The Organic Chemistry of Copper," *Adv. Organomet. Chem.,* 1974, **12,** 215.

Kauffman, T., "Oxidative Coupling of Organocopper Compounds," *Angew. Chem., Int. Ed. Engl.,* 1974, **13,** 291.

Bacon, R. G. R. and Hill, H. A. O., "Copper-Promoted Reactions in Aromatic Chemistry," *Q. Rev.,* 1965, **19,** 95.

Fanta, P. E., "The Ullmann Synthesis of Biaryls," *Synthesis,* **1974,** 9.

Preparation of Organocopper Compounds from Organic Halides and Highly Reactive Copper Powder

Rieke, R. D. and Rhyne, L. D., *J. Org. Chem.,* 1979, **44,** 3445.

Preparation of Organocopper Compounds from Alkenylboranes

Campbell, J. B. Jr. and Brown, H. C., *J. Org. Chem.,* 1980, **45,** 549, 550.

2

Organocopper Addition Reactions

A. Acyclic Ethylenic Ketones

Synthesis 1: ar-Turmerone,* A Spice and Dye Constituent

Spices, dyes, and perfumes extracted from plants have been used for many thousands of years. Even today the food, dye, and perfume industries rely heavily on plants as sources of flavors, colors, and scents which are added to much of our food, clothing, and cosmetics. Turmeric, an herb of the ginger family indigenous to southern Asia, was listed as a coloring plant in an Assyrian herbal dating from about 600 B.C. Currently, over 100,000 tons of cured Indian turmeric are produced annually; ground turmeric gives color and flavor to prepared mustard, curries, margarine, butter, and cheese, and it is included in many pickle and relish formulas. In Asia it is used as a medicinal agent and as a cosmetic. The essential oil of turmeric contains a high percentage of ketones, especially ar-turmerone.

The modern food, dye, and perfume industries are deeply involved in the basic organic chemistry of obtaining *pure constituents* from crude plant extracts. Such pure constituents are needed for evaluation

---

*Here "ar" means "aromatic"; turmerone itself is the corresponding nonaromatic 1,4-cyclohexadiene ketone.

of the taste, color, and odor of each constituent individually and in known combinations with other *pure* organic compounds. The art of evaluating the taste and the smell of a chemical is still in its infancy; even such structurally similar compounds as *cis-trans* isomers and *d*- and *l*-enantiomers may have very different tastes and odors. One important goal is to mix pure substances in order to prepare in the laboratory spices, dyes, or perfumes indistinguishable from, or even more desirable than, the natural ones. A major benefit associated with such manufactured spices, dyes, or perfumes is the "quality control" with which they can be prepared, in contrast to the often variable cost and chemical composition (and therefore variable taste, color, and odor) of natural plant extracts. When isolation of a valuable constituent of a plant extract is difficult or when the desired component is present in insufficient amounts, then laboratory *synthesis* of the constituent may be helpful.

Many of the essential constituents of turmeric have been synthesized. *d,l*-ar-Turmerone (**1-1**) has been synthesized in several different ways, one of which involves use of an organocopper reagent. Let us consider for a moment the chemical structure of ar-turmerone (**1-1**) and some general comments about organic synthesis. Take a few minutes to consider how *you* would approach synthesis of ar-turmerone!

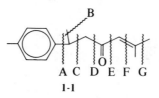

1-1

One reasonable assumption is that our synthesis should involve a preformed (commercially available or easily accessible) aromatic ring compound carrying two para substituents, one a methyl group and the other capable of undergoing attachment of carbon units in some fashion. Using this simplification, we can now examine the *most direct and general* possibilities for formation of carbon-carbon bonds A–G, indicated by the wavy lines in structure **1-1**.

The discussion that follows is a simplified version of retrosynthetic analysis, an approach made popular by Professor E. J. Corey of Harvard University. In this approach each carbon-carbon bond of a target mole-

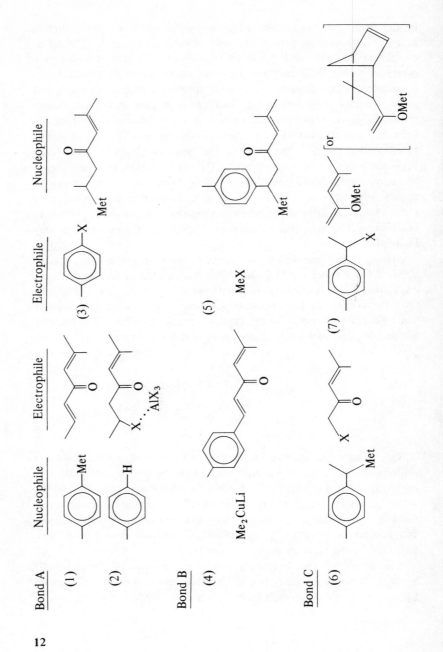

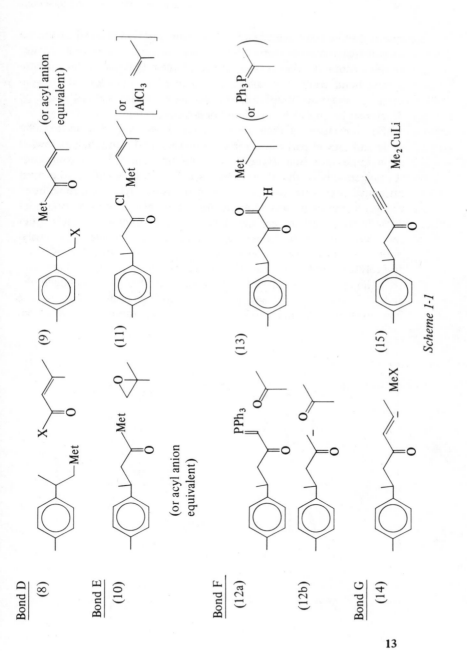

Scheme 1-1

13

cule is disconnected and then the following question is asked: is there a known reaction for connection of the carbon-carbon bond under consideration? If the answer is "no," then that particular carbon-carbon bond connection might be discarded as a viable route to synthesis of the target molecule, or a new synthetic method might be developed for this carbon-carbon bond-forming step.

For formation of each carbon-carbon bond A–G we can consider ionic and radical pathways; for ar-turmerone (**1-1**), the ionic pathways seem more promising (scheme 1-1). Of the four means outlined for forming bonds A and B, approaches (1) and (4) would be the most attractive if it were possible to mask the isopropylidene double bond easily; fortunately, protection-deprotection of the double bonds of α,β-ethylenic ketones can be done via Diels Alder–retro Diels Alder reactions with cyclopentadiene as suggested in scheme 1-2. β-Tolyl enone **1-2**, R = tolyl, is preferred over β-methyl enone **1-2**, R = Me, for tolyl-enone **1-2** is easier to prepare; *p*-tolualdehyde, lacking α-protons, undergoes a cleaner aldol condensation than does acetaldehyde. Indeed, tolyl-enone **1-2** and a copper-catalyzed Grignard reagent have been used as shown in eq. 1-1 for a successful synthesis of ar-turmerone:

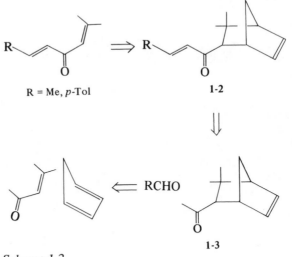

R = Me, *p*-Tol   **1-2**

1-3

Scheme 1-2

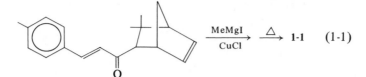

$$\text{(1-1)}$$

Inspection of the other possibilities in scheme 1-1 for formation of bonds C–G shows that only the direct approaches 7, 11, 12, and 15 look good. Specifically, route 7 has potential even though some bothersome side reactions could occur, such as β-elimination of H—X from the benzylic electrophile or equilibration of the kinetic enolate ion to the thermodynamically more stable dienolate $(CH_3COCH=C(CH_3)CH_2Met)$. Approach 11 is reasonable provided that the vinylmetallic reagent is sufficiently tame (e.g., vinylcopper species) so that it reacts with the acid chloride functionality but not with the ketone produced (i.e., not with ar-turmerone). Alternatively, Friedel-Crafts acylation of isobutylene might occur in competition with intramolecular Friedel-Crafts acylation (i.e., cyclization) of the β-aryl acyl chloride. Route 12a has potential; the required Wittig reagent is easily prepared from triphenylphosphine and the appropriate α-haloketone. Preparation of the α-haloketone (or of the corresponding enolate required in route 12b), however, occurs regio*selectively*, but probably not without formation of a small amount of the undesired α'-haloketone. Because the corresponding ketone is an available natural product (curcumone), approaches 12a and 12b take on additional significance. Approach 15 looks fine, provided that the acetylenic ketone is easily prepared.

What are the most troublesome aspects of the other approaches: 2, 3, 5, 6, 8, 9, 10, 13, and 14? The nucleophiles in approaches 3, 5, 8, 9, and 14 are hard to prepare. Likewise, the electrophiles in approaches 2 and probably 13 are hard to prepare. So we are left with possibility 6. Approach 6 suffers from very high reactivity of the electrophile; the desired carbon-carbon bond formation will probably not occur with the exclusion of other carbon-carbon connections.

A less direct but nevertheless interesting approach involves bond D connection, as outlined in eq. 1-2. Note the following two points: (1) the nucleophile is now a more accessible *allylic*-metallic species (cf. approach 8); (2) selective dissolving metal (e.g., Li/NH$_3$) reduction of an aryl-conjugated carbon-carbon double bond can be performed without concomitant reduction of an isolated carbon-carbon double bond.

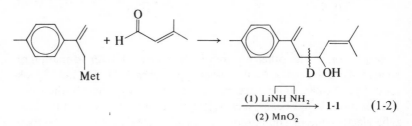

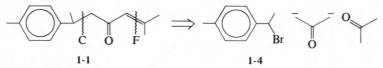

$$\xrightarrow[\text{(2) MnO}_2]{\text{(1) LiNH NH}_2} \text{1-1} \qquad (1\text{-}2)$$

Thus far we have analyzed the structure of ar-turmerone one bond at a time, considering reasonable methods for forming each bond separately. Suppose we were able to think of a way to prepare ar-turmerone by forming two bonds at one time or two bonds consecutively in the same reaction vessel. Such an approach would be highly efficient and creative. If we focus attention on the acetone portion of ar-turmerone, we see that bond C connects the acetone fragment with a benzylic group and bond F with an isopropylidene group (scheme 1-3); thus, the dianion of acetone (or its synthetic equivalent) should allow us to connect both bonds C and F. A β-keto phosphonate dianion was actually used in this sophisticated way to prepare ar-turmerone (eq. 1-3).

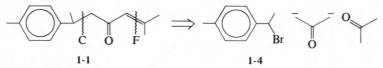

Scheme 1-3

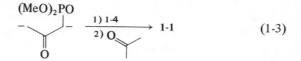

$$\xrightarrow[\text{2) O}]{\text{1) 1-4}} \text{1-1} \qquad (1\text{-}3)$$

References to Synthesis of ar-Turmerone

Rupe, H. and Gassmann, A., *Helv. Chim. Acta,* 1936, **19**, 569. (approach 12b)

Cologne, J. and Chambion, J., *C. R.,* 1946, **222**, 557. (approach 11)

Ghandi, R. P., Vig, O. P., and Mukherji, S. M., *Tetrahedron,* 1959, 7, 236. (modified approach 12b)

Howard, V. K. and Rao, A. S., *Tetrahedron,* 1964, **20,** 2921. (absolute configuration determined)

Crawford, R. J., Erman, W. F., and Broaddus, C. D., *J. Am. Chem. Soc.,* 1972, **94,** 4298. (modified approach 8)

Grieco, P. A. and Finkelhor, R. S., *J. Org. Chem.,* 1973, **38,** 2909. (combined approaches 7 and 12a)

Ho, T. L., *Synth. Commun.,* 1974, 4, 189. (approaches 4 and 7)

Park, O. S., Grillasca, Y., Garcia, G. A., and Maldonado, L. A., *Synth. Commun.,* 1977, 7, 345. (approaches 4 and 7)

Gosselin, P., Masson, S., and Thuillier, A., *J. Org. Chem.,* 1979, **44,** 2807. (approaches 8 and 11)

Meyers, A. I. and Smith, R. K., *Tetrahedron Lett.,* 1979, 2749. (approaches 1 and 12)

Masaki, Y., Hashimoto, K., Sakuma, K., and Kaji, K., *J. Chem. Soc., Chem. Commun.,* 1979, 855.

General Articles on Chemistry and Flavor

Chem. Soc. Rev., 1978, 7, 167–218.

Texts on the Strategy of Organic Synthesis

Warren, S., *Designing Organic Syntheses, A Programmed Introduction to the Synthon Approach,* John Wiley & Sons, Inc., New York, 1978.

Turner, S., *The Design of Organic Synthesis,* Elsevier Scientific Publishing Co., Amsterdam, 1976.

Fleming, I., *Selected Organic Syntheses, A Guidebook for Organic Chemists,* John Wiley & Sons, Inc., New York, 1972.

Ireland, R. E., *Organic Synthesis,* Prentice Hall, Inc., Englewood Cliffs, New Jersey, 1969.

Synthesis 2: 16α-Methyl-9α-fluoroprednisolone, An Anti-Inflammatory Steroid Drug

Steriods are crucial for the normal growth, development, and well-being of animals (including man) and possibly also of higher plants. Some examples of the kinds of steroids vital to humans include sex hormones, bile acids, cardiac glycosides, sapogenins, and adrenal cortical hormones. Perhaps the most generally known adrenal cortical hormone is cortisone, which is used medically for treatment of rheumatoid arthritis, asthma, and many skin disorders.

How does the pharmaceutical industry discover new drugs? Generally, new drugs are found in one of two ways: (1) by *screening* a very broad range of available chemicals for their physiological activity in animals; and (2) by *designing* new chemicals that may have the physiological effects being sought (animal testing is also done for these new compounds).

Rational design of new drugs is most effectively achieved by making relatively minor structural modifications in compounds whose physiological activity is high and whose mechanism of action and mode of deactivation (i.e., metabolic pathways) *in vivo* are known. For example, it was discovered accidentally that replacing the 9α-H by a 9α-F in various biologically active steroids often increases their potency; 9α-fluorohydrocortisone is 10 times more active than hydrocortisone in controlling carbohydrate metabolism but also is more active in undesirably raising salt concentrations in the body. Thus for a structural change in a compound to lead to a new drug, that structural change must ideally accentuate a compound's beneficial physiological effects without also increasing its harmful side effects.

Many corticosteroids possess the 17-hydroxy-20-keto (α-ketol) functionality, and normal human metabolism degrades these compounds into the corresponding 17-keto steroids, which have very different physiological properties (eq. 2-1). Thus a rational way to increase the lifetime (and therefore the effectiveness) of a corticosteroid drug is to make a structural change that retards metabolic cleavage of the α-ketol group. Placing a methyl group on carbon-16 causes the structural environment of the α-ketol group to be more crowded, and therefore normal biological oxidative cleavage of the side chain is retarded.

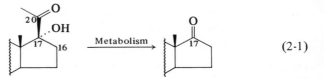

$$(2\text{-}1)$$

16α-Methyl-9α-fluoroprednisolone (Dexamethasone, **2-1**), for example, combines the potentiating effect of the 9α-F substituent with the longevity effect of the 16-methyl group; dexamethasone not only has a longer lifetime *in vivo* than 9α-fluoroprednisolone, but unexpectedly the additional 16-methyl group completely counteracts the strong salt-retaining effect of placing a fluorine atom at carbon-9 of prednisolone. Dexamethasone currently is a synthetic commercial drug used as a more potent, longer-lived, and less harmful substitute for cortisone.

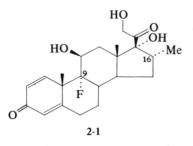

2-1

How is a 16-methyl group introduced into a 17-hydroxy-20-keto-corticosteroid? Let us consider how to introduce the 16-methyl and the 17-hydroxyl groups first independently and then together. α-Hydroxy ketones can be formed via epoxidation and subsequent basic hydrolysis of enol acetates. Noticing that a 16-methyl group is attached to a carbon atom that is β to a ketone carbonyl group, we come up with the retrosynthetic analysis shown in scheme 2-1, the first synthetic step of which involves β-addition of a methylcopper species to an α,β-ethylenic ketone. Notice that the organocopper conjugate addition to ethylenic ketone **2-3** does not disturb the 11-keto group or the homoallylic acetate functionality in rings A and B; such chemospecificity is one very useful characteristic of organocopper reagents.

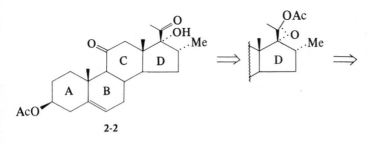

2-2

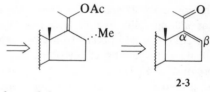

2-3

Scheme 2-1

Introducing the 16-methyl and 17-hydroxyl groups *together* might be a significant improvement over scheme 2-1. One possibility is to oxidize the enolate intermediate generated by the organocopper conjugate addition directly with lead tetracetate or with a molybdenum peroxide (eq. 2-2). Attachment of the 16-methyl and the 17-hydroxyl groups probably occurs with the desired stereochemical orientation due to the directing influence of the angular 18-methyl group.

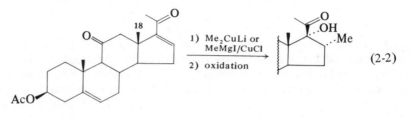

(2-2)

A second possibility for simultaneous elaboration of the 16-methyl and 17-hydroxyl groups is to perform an organocopper nucleophilic opening of an epoxide (eq. 2-3a). There are three serious problems with this approach: (1) the *trans*-opening of the epoxide via backside attack would lead to a 16β-methyl orientation; (2) nucleophilic attack is more favorable electronically at the undesired carbon-17 position (α to the carbonyl), although steric factors would favor carbon-16 attack; and (3) although α,β-epoxy *esters* undergo nucleophilic opening with organocopper reagents, α,β-epoxy *ketones* undergo overall reduction of the C_α-epoxide bond to form a β-hydroxyketone (eq. 2-3b). Indeed, most ketones bearing a good leaving group on the adjacent carbon undergo dimethylcopperlithium-promoted reductive cleavage, presumably via intermediacy of the corresponding enolates, as is the case in comparable dissolving metal or zinc promoted reactions (eq. 2-4).

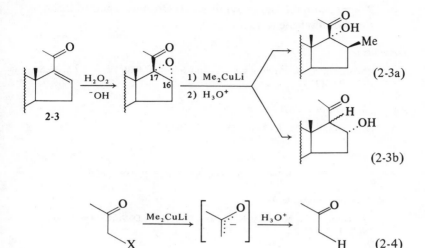

(2-3a)

(2-3b)

(2-4)

Conversion of steroidal ketone **2-2** all the way to 16α-methyl-9α-fluoroprednisolone (**2-1**) follows established routes that are not discussed here (see references below).

References to Synthesis of 16-Methylcorticosteroids

Arth, G. E., Johnston, D. B. R., Fried, J., Spooncer, W. W., Hoff, D. R., and Sarett, L. H., *J. Am. Chem. Soc.,* 1958, **80,** 3060.

Arth, G. E., Fried, J., Johnston, D. B. R., Sarett, L. H., Silber, R. H., Stoerk, H. C., and Winter, C. A., *J. Am. Chem. Soc.,* 1958, **80,** 3061.

Olivetto, E. P., Rausser, R., Nussbaum, A. L., Gebert, W., Hershberg, E. B., Tolksdorf, S., Eisler, M., Perlman, P. L., and Pechet, M. M., *J. Am. Chem. Soc.,* 1958, **80,** 4428.

Heusler, K., Kebrle, J., Meystre, C., Uberwasser, H., Wieland, P., Anna, G., and Wettstein, A., *Helv. Chim. Acta,* 1959, **52,** 2043.

Weiss, M. J. and Schaub, R. E., *J. Med. Chem.,* 1967, **10,** 789.

Mori, H. and Oh-Uchi, R., *Chem. Pharm. Bull.,* 1975, **23,** 559.

Cairns, J., Hewett, C. L., Logan, R. T., McGarry, G., Stevenson, D. F. M., and Woods, G. F., *J. Chem. Soc., Perkin I,* 1976, 1558.

Recent Natural Product Synthesis via Organocopper Addition to Acyclic Ethylenic Ketones

Longifolene

> Volkmann, R. A., Andrews, G. C., and Johnson, W. S., *J. Am. Chem. Soc.*, 1975, **97**, 4777.

Antibiotic Erythronolide A

> Hanessian, S., *Can. J. Chem.*, 1978, **56**, 1843.

(±)-Epiuleine

> Natsume, M. and Kitagawa, Y., *Tetrahedron Lett.*, **1980**, 839.

Carbocyclic Thromboxane A$_2$

> Nicolaou, K. C., Magolda, R. L., and Claremon, D. A., *J. Am. Chem. Soc.* 1980, **102**, 1404.

Oxidation of Enolates to α-Ketols

Ellis, J. W., *Chem. Commun.*, **1970**, 406. (Pb(OAc)$_4$)

Vedejs, E., Engler, D. A., and Telschow, J. E., *J. Org. Chem.*, 1978, **43**, 188. (MoO$_5$ · Pyridine · HMPA)

Organocopper-promoted Reductive Cleavage of α-Heteroatom-Substituted Ketones

Bull, J. R. and Lachmann, H. H., *Tetrahedron Lett.*, **1973**, 3055.

Bull, J. R. and Tuinman, A., *Tetrahedron Lett.*, **1973**, 4349.

Posner, G. H. and Sterling, J. J., *J. Am. Chem. Soc.*, 1973, **95**, 3076.

Logusch, E. W., *Tetrahedron Lett.*, **1979**, 3365.

References for Further Reading

Lednicer, D. and Mitscher, L. D., *Organic Chemistry of Drug Synthesis*, Wiley-Interscience, New York, vol. I, 1977; vol. II, 1980.

Stern, E. S., Cavalla, J. F., and Price Jones, D., *The Chemist in Industry (2): Human Health and Plant Protection*, Clarendon Press, Oxford, England, 1974.

B. Acyclic Ethylenic Esters

Synthesis 3:   l-3-Phenylbutanoic Acid, An Optically Active Carboxylic Acid

Most substances found in nature occur in optically active form. Indeed, one enantiomer of a pair may have properties that are desirable and beneficial to man, whereas the mirror-image enantiomer may have no utility at all or may have properties that are undesirable or even harmful to man. For example, d-nootkatone, a bicyclic sesquiterpene that is one of the essential components of grapefruit, has a fruitlike odor and taste with a threshold value of 0.8 parts per million (ppm), whereas l-nootkatone has no fruitlike character at all and its threshold value is about 600 ppm.

Food additives and drugs in which one enantiomer is physiologically active and which today are sold as racemic mixtures may someday be required by the United States Food and Drug Administration to be sold only in their active form. Thus synthesis of optically active compounds may take on added practical importance.

Resolution of a racemic mixture into its component enantiomers is effective but wasteful because the undesired enantiomer is usually discarded. Preparation of only the desired enantiomer via a stereocontrolled synthesis is more desirable but also more difficult.

Let us look at some rational approaches to stereocontrolled synthesis of a simple optically active carboxylic acid; although l-3-phenylbutanoic acid has no practical value currently, analyzing its structure and designing asymmetric syntheses will help us appreciate some of the problems encountered in preparing enantiomerically pure compounds.

The chiral center is at carbon-3 of l-3-phenylbutanoic acid and, therefore, unlike a chiral center adjacent to a carbonyl group, it is not easily epimerized by acid or base. Take some time to consider how you would approach stereocontrolled synthesis of this relatively simple structure.

Before considering how to disconnect and reconnect carbon-carbon bonds, we can examine whether manipulation of any functional groups will lead to the desired optically active acid. 3-Phenyl-2-butenoic (β-methyl cinnamic) acid can be hydrogenated to form 3-phenylbutanoic

acid; can this reduction of the conjugated carbon-carbon double bond
be done with control of stereochemistry? Asymmetric catalytic hydrogenation of acrylic acids using homogeneous rhodium catalysts carrying
optically active ligands has led to optically active carboxylic acids (e.g.,
α-amino acids) with excellent optical purity. We expect eq. 3-1 to proceed also with good stereocontrol.

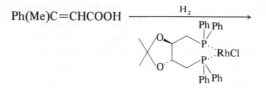

$$Ph(Me)\overset{*}{C}HCH_2COOH \quad (3\text{-}1)$$

Retrosynthetic analysis of the carbon skeleton of 3-phenylbutanoic
acid one bond at a time leads to the direct approaches shown in scheme
3-1.

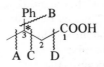

Approach 1 involves nucleophilic conjugate addition of an organometallic reagent (R-Met) to crotonic (R' = Me) or cinnamic (R' = Ph)
acid or to the corresponding esters. Organocopper reagents are generally
very effective for such conjugate additions to acyclic ethylenic esters.
Our problem, however, is not only to construct the correct carbon
skeleton but also to produce only one of two possible enantiomers.
There are several ways to use approach 1 for asymmetric synthesis of
l-3-phenylbutanoic acid: (1) use of an optically active R'' group in the
ester $RC\overset{\beta}{H}=C\overset{\alpha}{H}COOR''$ (e.g., R'' = menthyl), anticipating transfer of
chirality from the R'' group to the β-carbon as the conjugate addition
proceeds; (2) use of an optically active group R' in a mixed homocuprate
R(R')CuLi or an optically active group Z in heterocuprates R(Z)CuLi;
and (3) use of an optically active solvent. All of these three procedures
that involve asymmetric induction have been tried with only very limited
success (i.e., enantiomeric purity < 10%). The recent use of RCu · BR$_3$'

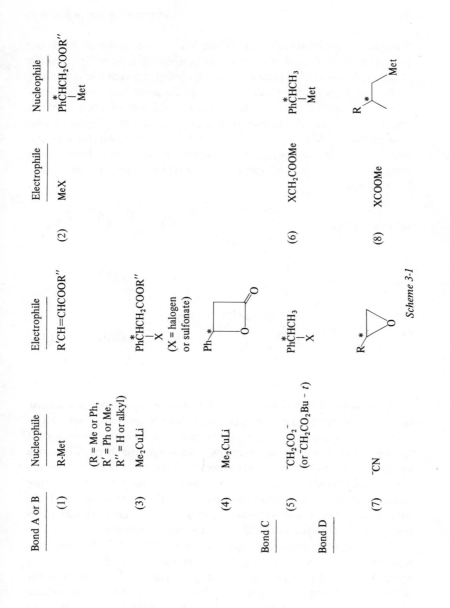

Scheme 3-1

reagents for conjugate addition to ethylenic esters, however, holds out some promise that $RCu \cdot BR_3'$ reagents in which R' is an optically active group might lead to conjugate adducts with high asymmetric induction.

Recently, great success (i.e., > 90% optical purity) has been achieved in stereocontrolled nucleophilic conjugate addition of organometallic reagents to chiral α,β-ethylenic imines in which the chiral center is part of a chelating bidendate ligand (eqs. 3-2 and 3-3). Application of this sequence to construction of *l*-3-phenylbutanoic acid from either acetaldehyde or benzaldehyde or from crotonaldehyde or cinnamaldehyde is clear.

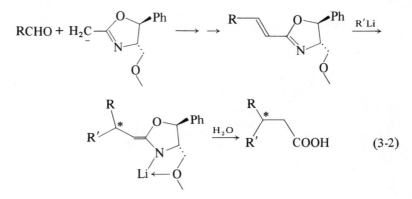

$$(3\text{-}2)$$

Approaches 3 and 4 in scheme 3-1 involve organocopper substitution at an activated (i.e., benzylic) secondary carbon atom with inversion of configuration at that center (see pp. 5–6 for discussion of organocopper substitution reactions). Although the benzylic leaving group is also β to the carboxyl carbonyl group, organocopper reagents are not sufficiently basic to cause α-deprotonation and therefore elimination of H—X to form cinnamic acid. The most serious problem with approaches 3 and 4, however, may be the relative difficulty in preparing the required optically active benzylic derivatives.

The only other reasonable approach in scheme 3-1 is that involving formation of bond C via attack of a two-carbon nucleophile on an optically active 1-phenyl-1-haloethane (approach 5). Ester enolates (or the dianion of acetic acid) and optically active 1-phenyl-1-chloroethane are expected to give optically active 3-phenylbutanoic acid in good chemical and optical yields.

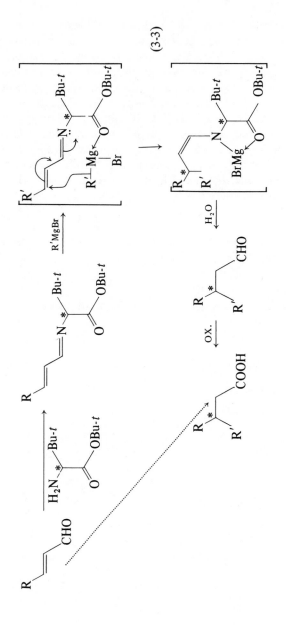

(3-3)

27

Approaches 7, 8, 6, and 2 have serious drawbacks. Although approach 7 involves initial nucleophilic opening of an unsymmetrical epoxide by cyanide ion, which occurs regioselectively at the less substituted position, the resulting β-hydroxynitrile probably cannot be hydrolyzed cleanly to the corresponding β-hydroxy carboxylic acid without significant dehydration; the initially formed β-hydroxynitrile itself, however, might be used as in approach 3 with nitrile hydrolysis delayed until the β-hydroxyl group has been replaced by a carbon substituent. Approach 8 suffers from the great difficulty in preparing the required optically active nucleophile. Approaches 6 and 2 both involve organometallic species in which the metal is attached to a chiral carbon atom; such species racemize very rapidly.

Recent Natural Product Syntheses via Organocopper Addition to Acyclic Ethylenic Esters

Fungal Antigerminating Agent Canadensolide

> Michihara, K., Kageyama, M., Tanaka, R., Kuwahara, K., and Yoshikoshi, A., *J. Org. Chem.,* 1975, **40**, 1932.

(+)-Hinesol

> Buddhsukh, D. and Magnus, P., *J. Chem. Soc., Chem. Commun.*, 1975, 952.

Vasoregulator Alkaloid Eburnamonine

> Casterousse, G., et al., *Bull. Soc. Chim. Fr.*, 1978, 355.

Boll Weevil Sex Attractant Grandisol

> Clark, R. D., *Synth. Commun.*, 1979, **9**, 325.

References for Further Reading

Asymmetric Induction

> Morrison, J. D. and Mosher, H. S., *Asymmetric Organic Reactions,* Prentice-Hall, Englewood Cliffs, New Jersey, 1971.
> Scott, J. W. and Valentine, D., Jr., *Science,* 1974, **184**, 943.

Homogeneous Asymmetric Catalytic Hydrogenation

> Kagan, H. B. and Dang, T. P., *J. Am. Chem. Soc.,* 1972, **94**, 6429.
> Dang, T. P. and Kagan, H. B., *J. Chem. Soc., Chem. Commun.,* 1971, 481.

Heterogeneous Asymmetric Catalytic Hydrogenation

> Takaishi, N., Imai, H., Bertelo, C., and Stille, J. K., *J. Am. Chem. Soc.,*
> 1978, **100**, 264.

R(Z chiral)CuLi Reagents

> Gustafsson, B., *Tetrahedron,* 1978, **34**, 3023.
> Ghozland, F., Luche, J. L., and Crabbé, P., *Bull. Soc. Chim. Belg.,* 1978,
> **87**, 369. See also Imamoto, T. and Mukaiyama, T., *Chem. Lett.,* 1980, 45.

Organocopper Addition to Chiral Acrylate Esters

> Asami, M. and Mukaiyama, T., *Chem. Lett.,* 1979, 569.

RCu · BF₃

> Yamamoto, Y. and Maruyama, K., *J. Am. Chem. Soc.,* 1978, **100**, 3240.

Organometallic Conjugate Addition to Ethylenic Imines

> Meyers, A. I. and Whitten, C. E., *J. Am. Chem. Soc.,* 1975, **97**, 6266.
> Meyers, A. I., Smith, R. K., and Whitten, C. E., *J. Org. Chem.,* 1979, **44**,
> 2250.
> Hashimoto, S., Yamada, S., and Koga, K., *J. Am. Chem. Soc.,* 1976, **98**,
> 7450.

Organometallic Reactions of Ester Enolates

> Rathke, M. W. and Lindert, A., *J. Am. Chem. Soc.,* 1971, **93**, 2318.
> Rathke, M. W. and Sullivan, D. F., *J. Am. Chem. Soc.,* 1973, **95**, 3050.
> Bergelson, L. D. and Shemyakin, M. M., in Patai, S., *The Chemistry of
> Carboxylic Acids and Esters,* Wiley-Interscience, New York, 1969.

C. Cyclic Ethylenic Ketones

Synthesis 4: β-Vetivone, A natural fragrance

β-Vetivone **(4-1)**, the major ketonic component of vetiver oil, has a
mild, sweet, and floral fragrance of interest to the perfume industry.

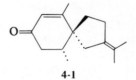

4-1

Preparation of racemic β-vetivone via an organocopper conjugate addition to cyclohexadienone **4-2** has been achieved. Although there are two possible sites for organocopper conjugate addition to dienone **4-2** (R = H), the less substituted β-position is attacked exclusively, as expected (eq. 4-1). Only when R = CHO, however, was introduction of the methyl group achieved stereoselectively (5 : 1 α:β orientation); an intermediate complex of type **4-2a** involving the formyl group and the isopropylidine double bond was invoked to account for this stereoselectivity.

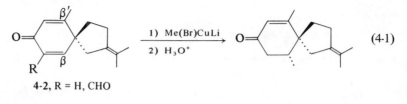

4-2, R = H, CHO   (4-1)

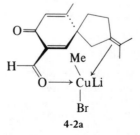

4-2a

The spiro-bicyclic skeleton of β-vetivone affords us an opportunity to analyze a fairly complex organic molecule with the aim of designing rational approaches to its synthesis. Let us consider the *general* problem of constructing the spiro[4.5]decane system characteristic of β-vetivone; consider how you would analyze this problem.

In this general retrosynthetic analysis we focus attention on the carbon atom common to the five- and six-membered rings (i.e., the spiro carbon atom). We ask first how to attach a five-membered ring in a spiro fashion to the same carbon atom of a preformed six-membered ring (six-membered rings are very common and relatively easy to prepare). Considering the organocopper route shown in eq. 4-1, we arrive at retrosynthetic scheme 4-1 involving cross-conjugated cyclohexadienone **4-3**.

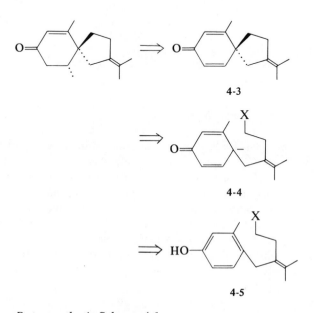

4-3

4-4

4-5

Retrosynthetic Scheme 4-1

Intramolecular displacement of leaving group X by the nucleophilic para carbon atom of *phenolate* **4-4** leads to formation of spiro[4.5]-decane **4-3**. Phenol **4-5** is therefore a logical and desirable synthetic intermediate that must now be analyzed. (Analysis of phenol **4-5** is left as an exercise for you).

Considering another albeit similar way to attach a five-membered ring in a spiro-fashion to a preformed six-membered ring, we arrive at retrosynthetic scheme 4-2.

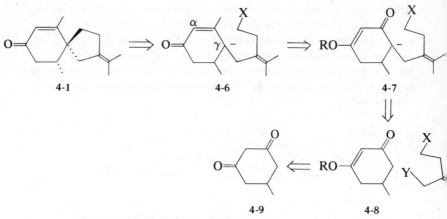

4-1   4-6   4-7

4-9   4-8

Retrosynthetic Scheme 4-2

Intramolecular displacement of leaving group X by the nucleophilic carbon atom of dienolate **4-6** produces β-vetivone with the proper relative stereochemistry. To prevent dienolate **4-6** from reacting also at its nucleophilic α-position to form an undesired bridged-bicyclic system, cross-conjugated dienolate **4-7** is chosen, since transformation 4-2 is easily carried out and dienolate **4-7** can probably be prepared via kinetic deprotonation of its neutral conjugate acid. Alternatively, Lewis acid–promoted intramolecular cyclization of olefinic chloride **4-6a** also leads directly to β-vetivone.

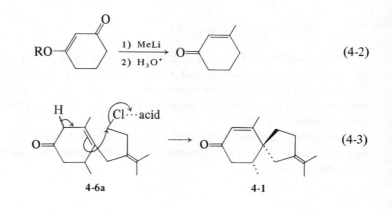

(4-2)

4-6a   4-1

(4-3)

How can we attach a six-membered ring in a spiro-fashion to a preformed five-membered ring? A straightforward retrosynthetic analysis
(scheme 4-3) disconnects the cyclohexenone ring into a 1,5-diketone
(which would undergo an aldol condensation in the synthetic direction)
and subsequently into a ketone and a vinyl ketone (which would
undergo a Michael addition); nonstereocontrolled synthesis of cyclohexenone **4-1** is accomplished using a Robinson annelation sequence
(**4-11** → **4-10** → **4-1**). A serious problem with this sequence is the
lack of control over which of the two acetyl groups in 1,5-diketone
4-10 acts as the electrophile and which as the nucleophile in an aldol
condensation.

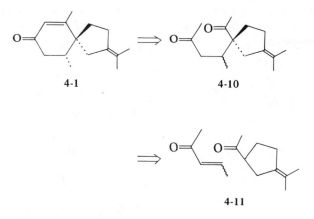

4-1   **4-10**

4-11

Retrosynthetic Scheme 4-3

Therefore an alternative approach involves elaborating the cyclohexenone ring of β-vetivone (**4-1**) via Lewis acid–promoted cyclization of 5-hexenoyl chloride **4-12** (scheme 4-4).

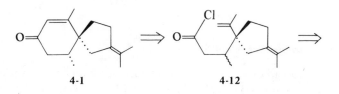

4-1   **4-12**

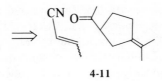

4-11

Retrosynthetic Scheme 4-4

To complete our general analysis, we ask whether there are any methods for converting *fused* bicyclic into *spiro* bicyclic systems. Many carbocyclic ring expansions and ring contractions are known, and the biosynthesis of spiro[4.5]decanes is thought to proceed via carbonium ion–induced Wagner-Meerwein rearrangement of a bicyclic decalin system into a spiro[4.5]decane. We therefore arrive at retrosynthetic scheme 4-5, patterned on the proposed biogenesis of these spiro compounds.

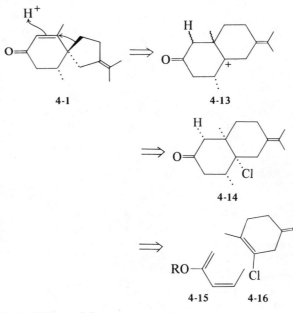

4-1   **4-13**

4-14

4-15   **4-16**

Retrosynthetic Scheme 4-5

Tertiary carbonium ion **4-13** can rearrange with loss of a proton to form β-vetivone (**4-1**). Carbonium ion **4-13**, however, can also undergo an undesired 1,2-methyl shift with proton loss to form *vicinal*-dimethyl-decalin **4-17** (eq. 4-4). To prevent this undesired 1,2-methyl shift, *cis*-decalyl chloride **4-14** is chosen so that a concerted *syn*-1,3,-elimination of HCl with concomitant decalin→spiro[4.5]decane rearrangement can occur as shown by the arrows in structure **4-14a**. Vinylic chloride **4-16** is expected to undergo a polar Diels-Alder cycloaddition to yield *cis*-decalyl chloride **4-14** directly.

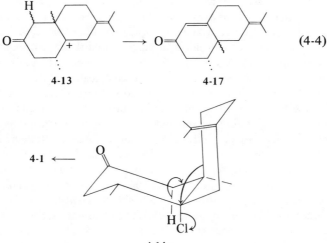

$$4\text{-}13 \qquad\longrightarrow\qquad 4\text{-}17 \qquad\qquad (4\text{-}4)$$

4-1 ⟵

4-14a

Synthesis of β-Vetivone Involving Organocopper Conjugate Addition

Mukharji, P. C. and Sen Gupta, P. K., *Chem. Ind.*, 1970, 533.

Bozzato, G., Bachmann, J. P., and Pesaro, M., *J. Chem. Soc., Chem. Commun.*, 1974, 1005.

Torii, S., Uneyama, K., and Okamoto, K., *Bull. Chem. Soc. Jpn.*, 1978, **51**, 3590.

A Recent Synthesis of β-Vetivone Based on Scheme 4-2

Johnson, A. P. and Vajs, V., *J. Chem. Soc., Chem. Commun.*, 1979, 817.

An Excellent Review with Many References to the Chemistry and Synthesis of Spiro[4.5]decane Sesquiterpenes

Marshall, J. A., Brady, S. F., and Andersen, N. H., *Progr. Chem. Org. Nat. Prod.*, 1974, 31, 283.

Synthesis 5: Prostaglandins

Isolated in 1957 from sheep prostate glands but now known to be present in many human tissues, fluids, and organs, the prostaglandins were early recognized to possess very powerful effects on functioning of the human respiratory, gastric, digestive, renal, reproductive, nervous, endocrine, and cardiovascular systems. During the 1960s an enormous effort was started, especially by academic and pharmaceutical chemists, to synthesize natural and structurally modified prostaglandins in the development of new drugs. It was anticipated that such new prostaglandin drugs might be as significant a medical breakthrough as the antibiotics were in the 1940s. By the early 1970s there was an impressive synthetic methodology built for laboratory construction of cyclopentanoid prostaglandin systems, but it was becoming clear that the prostaglandins are broadly active in affecting so many physiological functions that they might not lead to the very basic medicinal breakthrough originally anticipated. Nevertheless, some therapeutic uses of the prostaglandins have been made; one type of prostaglandin ($PGF_{2\alpha}$), for example, is currently used as an abortifacient for inducing labor and terminating pregnancy during the second trimester, and another type is used to treat stomach ulcers. By 1974 the cover of one issue of *Chemical and Engineering News* boldly proclaimed the end of any chemical contribution to the prostaglandin area—"Prostaglandins: chemistry in hand, it's up to the biologists." Shortly thereafter, however, short-lived thromboxanes and prostacyclin, close structural relatives of the prostaglandins, were discovered to have dramatic effects on blood platelet aggregation (i.e., blood clotting and blood clot lysis). This discovery stimulated renewed chemical activity in preparing more stable and yet physiologically active analogs of the thromboxanes and especially of prostacyclin. In 1979 discovery of the structurally related leukotrienes and their involvement in the body's allergic responses stimulated additional interest in prostaglandlike substances. Undoubtedly many new chemical and biological discoveries and breakthroughs in this area will be made in the 1980s. The biogenesis of the prostaglandins, the thromboxanes, and prostacyclin from the same fatty acid precursor, arachidonic acid, is shown in scheme 5-1.

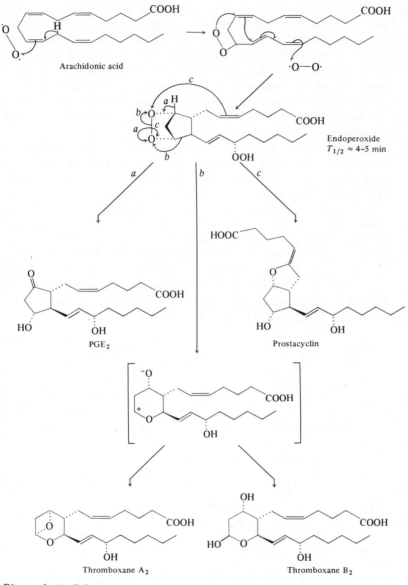

Arachidonic acid

Endoperoxide
$T_{1/2} \approx 4\text{--}5$ min

PGE$_2$

Prostacyclin

Thromboxane A$_2$

Thromboxane B$_2$

Biosynthetic Scheme 5-1

37

Several books chronicle the great chemical achievements in prostaglandin synthesis since 1965 (specific references are provided at the end
of this chapter). As an introduction to synthesis, however, we focus our
attention here on the general rationales used in designing approaches
to the prostaglandin carbon skeleton; only for those approaches using
organocopper reagents do we focus in detail on the individual reactions.
We first consider approaches that do not use organocopper reagents,
and second those that do. What rational approaches would you design
for construction of the prostaglandins?

Corey's brilliant retrosynthetic analysis of this problem is shown in
scheme 5-2. The two carbon-carbon double bonds in $PGF_{2\alpha}$ are formed
via Wittig and Horner-Wittig reactions with the corresponding aldehydes.
One of the aldehyde groups of dialdehyde **5-1** is masked as a lactone
(**5-2**, Corey's lactone), which allows sequential formation of the two
carbon-carbon double bonds. Aldehyde-lactone **5-2**, which has been
converted into all of the six primary prostaglandins ($PGE_{1,2,3}$, and
$PGF_{1\alpha,2\alpha,3\alpha}$), is prepared via iodolactonization and subsequent reductive deiodination (using R_3SnH) of olefinic acid **5-3**. The critical
stereochemical relationships in trisubstituted cyclopentene **5-3** are
formed in a completely controlled fashion from bicyclo[2.2.1]hepte-
none **5-5**, which is available via a Diels-Alder reaction.

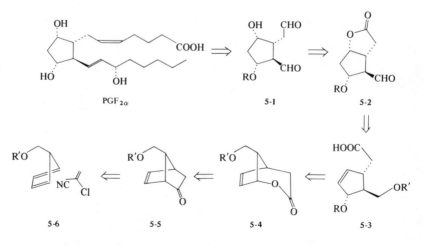

Corey Retrosynthesis: Scheme 5-2

Fried recognized that the *trans*-relationship between substituents at carbons-11 and -12 in cyclopentanol **5-7** (scheme 5-3) can be formed via nucleophilic opening of cyclopentene epoxide **5-9**, which carries an unprotected primary alcohol group to coordinate with and direct attack of the nucleophilic reagent toward carbon-12. Lithium diallylcuprate opens symmetrical cyclopentene epoxide **5-10** in a stereocontrolled *trans*-fashion to form a precursor to cyclopentene epoxide **5-9**.

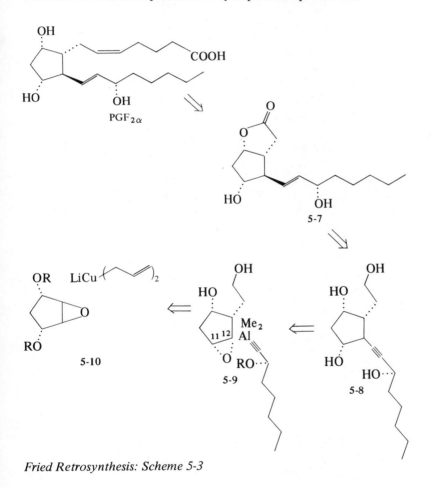

Fried Retrosynthesis: Scheme 5-3

Woodward's approach to the Corey lactone involves as the key carbon-carbon bond-forming step conversion of a six-membered ring into the required five-membered ring via an acid-catalyzed rearrangement of a *trans*-2-aminocyclohexanol (scheme 5-4). This approach is characterized by masterful manipulation of functional groups and consummate control of cyclohexane conformations.

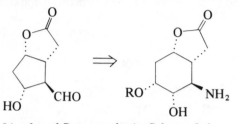

Woodward Retrosynthesis: Scheme 5-4

The key element in the Just-Upjohn retrosynthetic analysis involves the homoallylic alcohol function at carbons-11–14 of $PGF_{1\alpha}$ (scheme 5-5). Solvolysis of cyclopropylcarbinyl sulfonates leads generally to homoallylic products, and therefore cyclopropylcarbinyl mesylate **5-10** is chosen as a direct precursor to $PGF_{1\alpha}$ methyl ester.

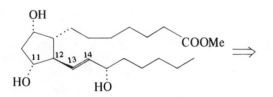

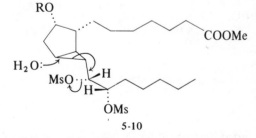

5-10

Upjohn-Just Retrosynthesis: Scheme 5-5

We turn now to organocopper conjugate addition to 2-cyclopen-
tenones as the key carbon-carbon σ-bond-forming step in some very
successful and widely used prostaglandin syntheses. The Sih-Syntex
approach (scheme 5-6) involves conjugate addition of the eight-carbon
"lower" side chain as an optically active or racemic vinylic-copperlithium
reagent to a racemic or optically active 4-alkoxy-2-cyclopentenone.
The virtually complete stereocontrol exhibited in this addition reaction
is typical of organocopper chemistry: the new bond to carbon-12 is
formed *trans* to the C_{11}-alkoxy group, and the functionalized vinylic
side chain is transferred from copper to carbon-12 with retention of
double bond geometry. Furthermore, this organocopper reaction
proceeds chemospecifically without consuming the methyl ester functionality, and protonation of the enolate intermediate generated via
this organocopper conjugate addition leads exclusively to the thermodynamically stable *trans*-orientation of the two adjacent carbon side
chains. Because the eight-carbon "lower" side chain is such a valuable
fragment, use is made of new mixed homocuprates R(R')CuLi and of the
R(Z)CuLi heterocuprates in which only R is transferred instead of using
R_2CuLi, which usually wastes one of the two R groups; the most
effective R(R')CuLi and R(Z)CuLi reagents are those with R = R'C≡C
and Z = PhS and *t*-BuO. The convergent nature and the efficiency of
this approach may soon allow practical and commercial production of
some prostaglandins via this type of organocopper conjugate addition
process.

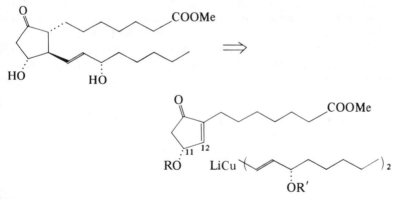

Sih-Syntex Retrosynthesis: Scheme 5-6

It is fortunate that the Sih-Syntex approach utilized an *ether* rather than an *ester* protecting group for the carbon-11 hydroxyl function. α,β-Ethylenic ketones with good leaving groups (e.g., carboxylate, halide) in the γ-position are reduced by organocopperlithium reagents, whereas those with poorer leaving groups (e.g., alkoxide) in the γ-position undergo conjugate addition (eq. 5-1).

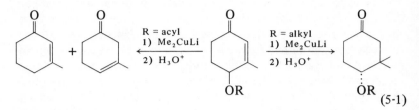

(5-1)

Likewise, ketones carrying adjacent carboxylate or halide groups are mainly reduced by organocopperlithium reagents, presumably via intermediacy of an enolate (eq. 5-2):

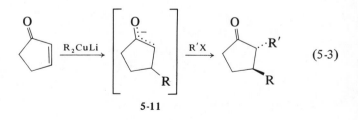

Enolate ions generated regiospecifically via organocopper β-addition to α,β-ethylenic ketones can be O-acylated and O-silylated and also C-alkylated; C-alkylation leads to an efficient method for introducing two different hydrocarbon groups (one nucleophilic and one electrophilic) β and α to a ketone, usually with a *trans* stereochemical relationship (eq. 5-3). This approach has thus far been used to construct prostaglandin model systems but not complete prostaglandin structures.

(5-3)

5-11

Although the nature of enolate intermediate **5-11** generated via organocopper conjugate addition is still controversial, its chemical reactivity is affected by the presence of copper and by what ligands are associated with copper. For example, mixed homocuprate R(R')CuLi and heterocuprate R(Z)CuLi reagents add to ethylenic ketones, producing enolate intermediates that react differently depending on the nature of R' and Z. Although successful enolate C-alkylations are usually achieved with R' = alkC≡C, more examples are needed before legitimate generalizations are possible; there is much experimentation still to be done in this area of organocopper-generated enolate chemistry.

Stork has developed the most effective and brilliant application of this type of cyclopentenone α,β-dialkylation procedure for prostaglandin synthesis. 4-Alkoxy-2-cyclopentenone **5-12** reacts with an eight-carbon vinylic cuprate and then with formaldehyde (i.e., an aldol condensation) to produce, after dehydration, a new α,β-ethylenic ketone which now reacts with a different functionalized vinylic cuprate to form PGE$_2$ directly (scheme 5-7).

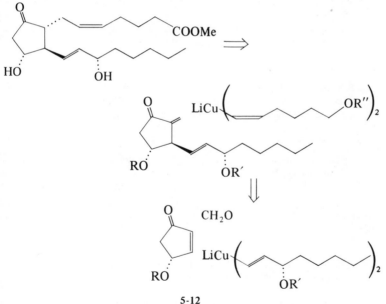

5-12

Stork Retrosynthesis: Scheme 5-7

Reviews of Prostaglandin Synthesis

Bindra, J. S. and Bindra, R., *Prostaglandin Synthesis,* Academic Press, New York, 1977.

Mitra, A., *The Synthesis of Prostaglandins,* Wiley-Interscience, New York, 1977.

Crabbé, P., Ed., *Prostaglandin Research,* Academic Press, New York, 1977.

An entire journal (*Prostaglandins,* founded in 1969) is devoted to prostaglandin research.

Corey Scheme 5-2

Corey, E. J., *Ann. N. Y. Acad. Sci.,* 1971, **180**, 24.

Fried Scheme 5-3

Fried, J. and Sih, J. C., *Tetrahedron Lett.,* **1973**, 3899.

Woodward Scheme 5-4

Woodward, R. B., Gasteli, J., Ernest, J., Friary, R. J., Nestler, G., Raman, H., Sitrin, R., Suter, C., and Whitesell, J. K., *J. Am. Chem. Soc.,* 1973, **95**, 6853.

For an overall view of the strategy used in this synthesis, see Ernest, I., *Angew. Chem., Int. Ed. Engl.,* 1976, **15**, 207.

Upjohn-Just Scheme 5-5

Schneider, W. P., Axen, U., Lincoln, F. H., Pike, J. E., and Thompson, J. L., *J. Am. Chem. Soc.,* 1968, **90**, 5895, and 1969, **91**, 5372.

Ferdinandi, E. S. and Just, G., *Can. J. Chem.,* 1971, **49**, 1070.

Sih-Syntex Scheme 5-6

Sih, C. J., Salomon, R. G., Price, P., Peruzzoti, G., and Good, R., *Chem. Commun.* 1972, 240.

Kluge, A. F., Untch, K. G., and Fried, J. H., *J. Am. Chem. Soc.,* 1972, **94**, 7827.

Stork Scheme 5-7

Stork, G. and Isobe, M., *J. Am. Chem. Soc.,* 1975, **97**, 6260.

(See also Davis, R. and Untch, K. G., *J. Org. Chem.,* 1979, **44**, 3755, for use of eq. 5-3 in prostaglandin synthesis.)

Organocopper Reaction with γ- and δ-Oxygenated Enones

(a) Hannah, D. and Smith, R., *Tetrahedron Lett.,* **1975**, 187; *Tetrahedron,* 1979, **35**, 1183.

(b) Ruden, R. A. and Litterer, W. E., *Tetrahedron Lett.*, 1975, 2043.

Organocopper Reaction with α-Halo and α-Acetoxy Ketones

(a) Posner, G. H. and Sterling, J. J., *J. Am. Chem. Soc.*, 1973, **95**, 3076;

(b) Bull, J. R. and Tuinman, A., *Tetrahedron Lett.*, 1973, 4349.

(c) Chuit, C., Sauvêtre, R., Masure, D., and Normant, J. F., *Tetrahedron*, 1979, **35**, 2645.

(d) Déprés, J.-P. and Greene, A. E., *J. Org. Chem.*, 1980, **45**, 2037.

References Since 1975 to Prostaglandin Syntheses Using Organocopper Additions to 2-Cyclopentenones

Bruhn, M., Brown, C. H., Collins, P. W., Palmer, J. R., Dajani, E. T., and Pappo, R., *Tetrahedron Lett.*, 1976, 235.

Woessner, W. D., Arndt, H. C., Biddlecom, W. G., and Peruzzotti, G. P., 172nd National A.C.S. Meeting, San Francisco, 1976, Abstract ORGN 9.

Wiel, J-P. and Rouessac, F., *J. Chem. Soc., Chem. Commun.*, 1976, 446.

Hallet, W. A., Wissner, A., Grudzinskas, C. V., and Weiss, M. J., *Chem. Lett.*, 1977, 51.

Kluender, H. C. and Peruzzotti, G. P., *Tetrahedron Lett.*, 1977, 2063.

Woessner, W. D., Arndt, H. C., Biddlecom, H. G., and Peruzzotti, G. P., 173rd National A.C.S. Meeting, New Orleans, 1977, Abstract ORGN 176.

Lüthy, C., Konstantin, P., and Untch, K. G., *J. Am. Chem. Soc.*, 1978, **100**, 6211.

Chen, S-M. L., Schaub, R. E., and Grudzinskas, C. V., *J. Org. Chem.*, 1978, **43**, 3450.

Hart, T. W., Metcalfe, D. A., and Scheinmann, F., *J. Chem. Soc., Chem. Commun.*, 1979, 156.

Schore, N. E., *Synth. Commun.*, 1979, **9**, 41.

Pernet, A. G., et. al., *Tetrahedron Lett.*, 1979, 3933.

Wiel, J-P. and Rouessac, F., *Bull. Soc. Chim. Fr.*, 1979, II-273.

References Since 1975 to Natural Products Prepared via Organocopper Addition to Cyclic Ethylenic Ketones

α- and β-Pinene

Thomas, M. T. and Fallis, A. G., *J. Am. Chem. Soc.*, 1976, **98**, 1227.

Podocarpic Acid

Huffman, J. W. and Harris, P. G., *J. Org. Chem.*, 1977, **42**, 2357.

Brefeldin A

Corey, E. J. and Wollenberg, R. H., *Tetrahedron Lett.*, **1976**, 4705;
Baudoy, R., Crabbé, P., Greene, A. E., Le Drian, C., and Ori, A. F., *Tetrahedron Lett.*, **1977**, 2973.

Solavetivone, a Stress Metabolite from Infected Potato Tubers

Yamada, K., Goto, S., Nagase, H., and Christensen, A. T., *J. Chem. Soc., Chem. Commun.*, **1977**, 554.

Eremophilone

Ficini, J. and Touzin, A. M., *Tetrahedron Lett.*, **1977**, 1081.

Acorone, a Spiro[4.5]decane

McCrae, D. A. and Dolby, L., *J. Org. Chem.*, **1977**, **42**, 1607.

Ishwarone

Cory, R. M. and McLaren, F. R., *J. Chem. Soc., Chem. Commun.*, **1977**, 587; and *Tetrahedron Lett.*, **1979**, 4133.

"Eastern Zone" of the Anticancer Compound Maytansine

Samson, M., DeClercq, P., DeWilde, H., and Vandewalle, M., *Tetrahedron Lett.*, **1977**, 3195.

Oplapanone

Taber, D. F. and Korsmeyer, R. W., *J. Org. Chem.*, **1978**, **43**, 4925.

δ-Lactones

Ficini, J., Kahn, P., Falou, S., and Touzin, A. M., *Tetrahedron Lett.*, **1979**, 67.

Prelog-Djerassi Lactone

White, J. D. and Fukuyama, Y., *J. Am. Chem. Soc.*, **1979**, **101**, 226.
Stork, G. and Nair, V., *J. Am. Chem. Soc.*, **1979**, **101**, 1315.

(±)-Gymnomitrol

Welch, S. C. and Chayabunjonglerd, S., *J. Am. Chem. Soc.*, **1979**, **101**, 6768.

Nagilactone

Hayashi, Y., Matsumoto, T., Hyono, T., Nishikawa, N., Uemura, M., Nishizawa, M., Togami, M., and Sakan, T., *Tetrahedron Lett.*, **1979**, 3311.

(±)-*Isocomene*

Paquette, L. A., and Han, Y. K., *J. Org. Chem.*, 1979, **44**, 4015.

Carvone

Fleming, I. and Paterson, I., *Synthesis*, 1979, 736.

Isolongifolene

Piers, E., and Zbozny, M., *Can. J. Chem.*, 1979, **57**, 2249.

Cuparene

Kametani, T., Tsubuki, M., and Nemoto, H., *J. Chem. Soc.*, Perkin I, 1980, 759.

References Since 1975 to Organocopper β-Addition and α-Alkylation of Cyclic α,β-Ethylenic Ketones

Greene, A. E. and Crabbé, P., *Tetrahedron Lett.*, 1976, 4867.

Toru, T., Kurozumi, S., Tanaka, T., Miura, S., Kobayashi, M., and Ishimoto, S., *Tetrahedron Lett.*, 1976, 4087.

Kurozumi, S., Toru, T., Tanaka, T., Kobayashi, M., Miura, S., and Ishimoto, S., *Tetrahedron Lett.*, 1976, 4091.

Tanaka, T., Kurozumi, S., Toru, T., Kobayashi, M., Miura, S., and Ishimoto, S., *Tetrahedron*, 1977, **33**, 1105.

Piers, E. and Lau, C. K., *Synth. Commun.*, 1977, **7**, 495.

Posner, G. H. and Lentz, C. M., *Tetrahedron Lett.*, 1977, 3215.

Posner, G. H., Whitten, C. E., Sterling, J. J., Brunelle, D. J., Lentz, C. M., Runquist, A. W., and Alexakis, A., *Ann. N.Y. Acad. Sci.*, 1978, **295**, 249.

Posner, G. H. and Lentz, C. M., *Tetrahedron Lett.*, 1978, 3769.

Semmelhack, M. F., Yamashita, A., Tomesch, J. C., and Hirotsu, K., *J. Am. Chem. Soc.*, 1978, **100**, 5565.

Posner, G. H. and Lentz, C. M., *J. Am. Chem. Soc.*, 1979, **101**, 934.

Garnero, J. and Joulain, D., *Bull. Soc. Chem. Fr.*, 1979, 15.

Boeckman, R. K., Jr., Blum, D. M., and Arthur, S. D., *J. Am. Chem. Soc.*, 1979, **101**, 5060.

Han, Y-K. and Paquette, L. A., *J. Org. Chem.*, 1979, **44**, 3731.

Posner, G. H., Chapdelaine, M. J., and Lentz, C. M., *J. Org. Chem.*, 1979, **44**, 3661.

Nicolaou, K. C. and Barnette, W. E., *J. Chem. Soc., Chem. Commun.* 1979, 1119.

Mixed Homocuprate R(R′)CuLi and Heterocuprate R(Z)CuLi Reagents

R′ = n-PrC≡C

Corey, E. J. and Beames, D. J., *J. Am. Chem. Soc.*, 1972, **94**, 7210.

$R' = t\text{-}BuC{\equiv}C$

House, H. O. and Umen, M. J., *J. Org. Chem.,* 1973, **38**, 3893.

$R' = Me_2NCH_2C{\equiv}C$

Ledlie, D. B. and Miller, G., *J. Org. Chem.,* 1979, **44**, 1006.

Z = Halogen

Luong-Thi, N. T. and Riviére, H., *Tetrahedron Lett.,* **1970**, 1583; **1971**, 587.

Z = PhS, t-BuO

Posner, G. H., Whitten, C. E., and Sterling, J. J., *J. Am. Chem. Soc.,* 1973, **95**, 7788.

Z = CN

Marino, J. P. and Floyd, D. M., *Tetrahedron Lett.,* **1979**, 675.

Synthesis 6: Vernolepin, An Anticancer Compound

Plants have provided us with spices, dyes, and fragrant compounds. Every culture in the world has also used plant preparations as medications to treat prevailing illnesses. Plant extracts have led to isolation of modern drugs such as quinine (an antimalarial), morphine (a pain killer), digitalis (a cardiac tonic and stimulant), cocaine (a surface anesthetic), and reserpine (a tranquilizer). The 1950s discovery that the trailing periwinkle plant, *Catharanthus roseus* G. Don, contains the alkaloids vinblastine and vincristine, which are now used as drugs in treating childhood leukemia, stimulated a worldwide search of plant extracts for potential anticancer drugs. This search, sponsored by the National Cancer Institute of the U.S. National Institute of Health, started with examination of the folk medicines of almost every primitive culture; also practitioners of indigenous medicine were contacted for help in locating appropriate plants. Of the first 40,000 plant extracts tested, approximately 3% had reproducible anticancer activity, and since the middle 1960s many more plants have been assayed. Biological assays have usually been done with two test systems: the KB tissue culture of human carcinoma of the nasopharynx, and the experimental P388 lymphocytic leukemia implanted in mice. Since 1970 over 250 natural products showing strong activity against these test systems have been isolated, and several of the most active and least toxic ones are currently undergoing human testing in chemotherapy clinics. Isolation of these potential anticancer drugs from plants is often a very difficult process, because most of these compounds represent very much less than 1% of the crude plant extracts, and they are usually accompanied by structurally similar but biologically inactive compounds. Continual biological assay of the fractions obtained during the separation procedure has been invaluable in allowing even minute quantities of biologically active compounds to be isolated.

Synthesis of anticancer compounds is important, not only because often only small quantities of these compounds are available from plant sources, but also because synthesis may lead to structural analogs having higher physiological activity, higher selectivity, and lower toxicity.

Alkylating agents are the most thoroughly studied class of antitumor compounds. Apparently, alkylating agents become covalently bound to DNA and often cross-link two neighboring strands of DNA, thus preventing replication and causing ultimate death of the cancerous cell. The high reactivity and low selectivity of most alkylating agents, however, often make them harmful to normal cells as well as to cancerous cells. Recently use has been made of latent alkylating species which require biochemical activation before they become alkylating agents; if cancerous cells are better able than normal cells to activate these latent alkylating agents, then cancerous cells can be selectively destroyed. Clearly, the differences between the biochemistry of cancerous and normal cells must be known before intelligent design of latent alkylating agents can be perfected.

In the late 1960s and early 1970s a variety of plant-derived antitumor sesquiterpene α-methylenelactone (i.e., Michael acceptor) alkylating agents were isolated. Among these, vernolepin (6-1), obtained in very small quantity, showed high antitumor activity. A sizable

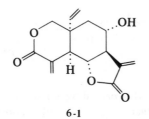

6-1

effort was mounted to synthesize vernolepin, and this effort culminated in several total syntheses reported in 1976. Whether vernolepin, or indeed any other plant-derived α-methylenelactone alkylating agent, eventually becomes a useful anticancer drug remains to be seen.

Although none of the reported total syntheses of vernolepin involves organocopper reagents, two syntheses of vernolepin prototypes depend on conjugate addition of a vinylcopper reagent for introduction of the characteristic angular vinyl group as well as the *cis*-ring junction in vernolepin. The first approach, represented in eq. 6-1, relies on organo-copper pseudoaxial conjugate addition followed by enolate trapping as the corresponding enol silyl ether. The second approach, albeit similar, involves a much more highly functionalized enone acceptor; copper-catalyzed vinylmagnesium bromide conjugate addition to ester ether cyclohexenone **6-2** proceeds chemospecifically (only the enone func-tionality reacts) and with high stereocontrol (>97.5% of the *cis*-ring junction is formed, eq. 6-2). A copper-catalyzed Grignard conjugate addition was used in this case mainly because vinyllithium was not commercially available and is much more difficult to prepare than is a vinyl Grignard reagent.

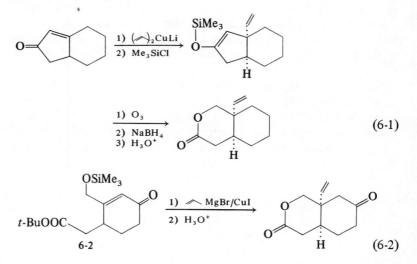

(6-1)

(6-2)

As emphasized in the preceding chapter, γ-acetoxy-α,β-ethylenic ketones react with organocopper reagents "abnormally" (i.e., with no carbon-carbon bond formation) to produce the corresponding α,β-ethylenic ketone; this reductive loss of the γ-acetoxy group is observed also in eq. 6-3. Mechanistically, this reaction can be viewed either as an electron transfer from the organocopper reagent to the enone (eq. 6-4) or as a cuprate addition-elimination reaction (6-5).

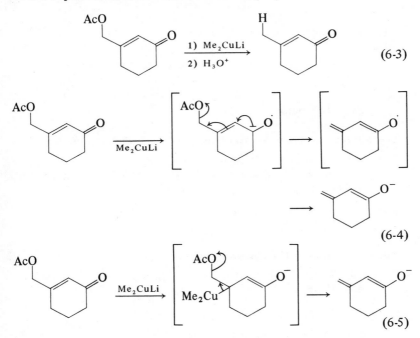

Vernolepin Prototype Syntheses using Organocopper Reagents

Clark, R. D. and Heathcock, C. H., *Tetrahedron Lett.*, 1974, 1713.

Wege, P. M., Clark, R. D., and Heathcock, C. H., *J. Org. Chem.*, 1976, 41, 3144.

Vernolepin Total Syntheses

Grieco, P. A., Nishizawa, M., Oguri, T., Burke, S. D., and Marinovic, N., *J. Am. Chem. Soc.,* 1977, 99, 5773.

Danishefsky, S., Schuda, P. F., Kitahara, T., and Etheredge, S. J., *J. Am. Chem. Soc.,* 1977, **99**, 6066.

Kieczykowski, G. P. and Schlessinger, R. H., *J. Am. Chem. Soc.,* 1978, **100**, 1938.

Iio, H., Isobe, M., Kawai, T., and Goto, T., *Tetrahedron,* 1979, **35**, 941.

Vinylcopper Conjugate Addition

Walliser, F. M. and Yates, P., *J. Chem. Soc., Chem. Commun.,* 1979, 1025.

Other References

Excellent Introductions to Natural Products Used in the Fight Against Cancer

Sainsbury, M., *Chem. Br.,* 1979, **15**, 127.

Sartorelli, A. C., Ed., *Cancer Chemotherapy,* ACS Symposium Series 30, Washington, D.C. 1976.

A Recent Study of α-Methylene-γ-lactone Inhibition of Tumor Growth

Hall, I. H., et al., *J. Pharm. Sci.,* 1978, **67**, 1235.

A Review of the Biological Activity of Sesquiterpene Lactones

Rodriguez, E., Towers, G. H. N., and Mitchell, J. C., *Phytochemistry,* 1976, **15**, 1573.

A Review of the Chemistry of Unsaturated Lactones

Rao, Y. S., *Chem. Rev.,* 1976, **76**, 625.

Synthesis 7: Lycopodine, An Alkaloid

Alkaloids are nitrogen-containing compounds that occur naturally in plants. About 15–20% of all vascular plants contain alkaloids, and more than 2000 alkaloids have been isolated. Many alkaloids are pharmacologically active. Examples include the following: quinine, morphine, codeine, cocaine, reserpine, lysergic acid diethylamide (LSD), belladonna (deadly nightshade), adrenalin, mescaline (a psychomimetic drug), and nicotine. Although lycopodine was first isolated in 1881, its full structure was not established until 1960. Several successful lycopodine total syntheses have been reported since 1967.

The most recent total synthesis of lycopodine involves a beautiful retrosynthetic analysis. Heathcock's creative insight into this problem allowed him to identify polycyclic intermediate **7-1** (scheme 7-1) as a β-amino ketone and therefore to design an intramolecular Mannich cyclization using 2,3,5-trisubstituted cyclohexanone **7-2**. The required *trans*-stereochemical relationship of the carbon-3 and carbon-5 substituents in 2,3,5-trisubstituted cyclohexanone **7-2** was obtained via conjugate addition of dimethallylcopperlithium or of

to a 2,5-disubstituted 2-cyclohexenone. The remarkably good stereocontrol in this organocopper conjugate addition reaction is typical for organocopper additions to cyclic enones which, based on stereoelectronic factors, lead almost exclusively to axial attachment of the new carbon-carbon bond.

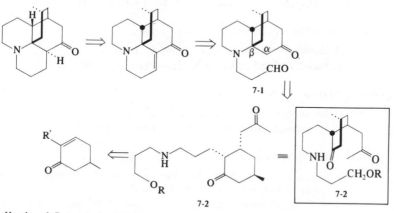

Heathcock Retrosynthetic Scheme 7-1

In passing, we note that copper enolates have been oxidatively dimerized into 1,4-dicarbonyl compounds and in some cases coupled with reactive (e.g., allylic) halides.

References for Further Reading

Lycopodine Synthesis Using Organocopper Reagents

Heathcock, C. H., Kleinman, E., and Binkley, E. S., *J. Am. Chem. Soc.*, 1978, **100**, 8036.

Reactions of α-Cuprioketones

Hampton, G. K. and Christie, J. J., *J. Org. Chem.*, 1975, **40**, 3887.

Dimerization of α-Cuprioketones

Kobayashi, Y., Taguchi, T., and Tokuno, E., *Tetrahedron Lett.,* 1977, 3471; Kobayashi, Y., Taguchi, T., Morikawa, T., Tokuno, E., and Sekiguchi, S., *Chem. Pharm. Bull.,* 1980, **28**, 262.

Reactions of α-Cuprioesters

(a) Corey, E. J. and Kuwajima, I., *Tetrahedron Lett.*, 1972, 487.

(b) Kuwajima, I. and Doi, Y., *Tetrahedron Lett.*, 1972, 1163.

(c) Rathke, M. W. and Lindert, A., *J. Am. Chem. Soc.*, 1971, **93**, 4605.

Reactions of α-Cupriophosphonates

(a) Matthey, F. and Savignac, P., *Synthesis*, 1976, 766;

(b) Varlet, J-M. and Collignon, N., *Synth. Commun.*, 1978, **8**, 335.

Reviews of the Mannich Reaction

(a) Blicke, F. F., *Org. React.*, 1942, **1**, 303.

(b) Hellmann, H. and Opitz, G., *Angew. Chem.*, 1956, **68**, 265.

Interesting and Relevant Books

Swain, T., Ed., *Plants in the Development of Modern Medicine*, Harvard University Press, Cambridge, Massachusetts, 1972.

Swan, G. A., *An Introduction to the Alkaloids*, John Wiley & Sons, New York, 1967.

D. Polyethylenic Carbonyl Compounds

Synthesis 8:   Ethyl (E)-2,(Z)-6-dodecanedioate, A Bartlett Pear Flavor

First isolated and characterized in 1970, the esters of (E)-2,(Z)-4-dodecanedioate and of (E)-2,(Z)-6-dodecanedioate are the major flavor components of Bartlett pears. Although the *conjugated* diolefinic esters are more abundant than the *nonconjugated* ones, both types are essential for construction of an artificial pear flavor resembling the natural one. This is one example of how flavor chemists use a combination of compounds to approximate the taste and smell of a natural food.

What approaches would you consider for construction of ethyl (E)-2,(Z)-6-dodecanedioate? Synthesis of ethyl (E)-2,(Z)-6-dodecanedioate (8-1) can be achieved in a variety of ways with varying degrees of chemical and stereochemical control. For example, the isolated *cis*-carbon-carbon double bond of nonconjugated diene 8-1 might be formed via a Wittig reaction using an unstabilized phosphorus ylide, and the conjugated *trans*-carbon-carbon double bond might be produced via a Wittig reaction using a stabilized acyl-phosphorane or via an aldol condensation using ethyl acetate anion or its synthetic equivalent (e.g., $LiCH_2COOLi$). Obviously the appropriate aldehyde partners in these Wittig and aldol condensations must be easily accessible if these approaches are to be viable. Alternatively, considering nonconjugated diene 8-1 as a 1,5-diene, we could imagine putting this molecule together via allylic coupling at the γ-carbon of ethyl crotonate with an allylic (i.e., (Z)-2-octenyl) halide; coupling at the α-carbon of ethyl crotonate, however, is probably the main carbon-carbon bond-forming reaction (eq. 8-1).

Optimum stereochemical control would be achieved during formation of the isolated carbon-carbon double bond via transfer of a nucleophilic *cis*-1-heptenyl group with retention of double bond geometry to an electrophilic pentenoate; retrosynthetic scheme 8-1 is generated in this way. Ethyl 2,4-pentadienoate reacts with *cis*-vinylcuprate reagents in a 1,6-addition fashion to give an intermediate enolate that is kinetically protonated α to the carbonyl group to form a β,γ-ethylenic ester

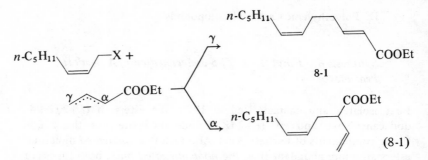

(8-1)

(eq. 8-2). Thermodynamic equilibration gives the desired α,β-ethylenic ester **8-1**.

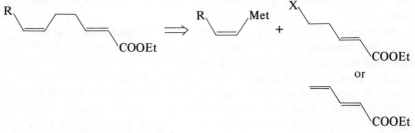

Retrosynthetic Scheme 8-1

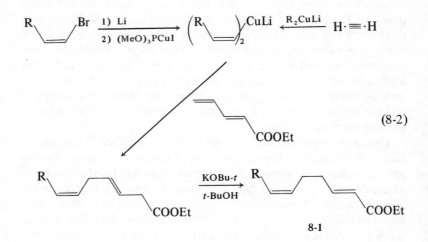

(8-2)

Preparation of the required *cis*-alkenyl cuprate reagent has been achieved in three ways: (1) via lithium metal reaction with the corresponding *cis*-vinylic bromide; (2) via bromine-lithium exchange using *t*-BuLi; and (3) via organocopper *cis*-addition to acetylene. The third method, first developed in France by J. F. Normant and co-workers, is a very direct means of preparing stereochemically homogeneous (Z)-alkenylcopper compounds that transfer the alkenyl group to various electrophiles with virtually complete retention of configuration.

References for Further Reading

Organocopper Approaches to (E)-2,(Z)-6-Dodecanedioate

Näf, F., Degen, P., and Ohloff, G., *Helv. Chim. Acta*, 1972, **55**, 82.

Alexakis, A., Normant, J., and Villiéras, J., *Tetrahedron Lett.*, **1976**, 3461.

(Z)-Alkenylcuprates via Organocopper Addition to Acetylenes

Alexakis, A., Cahiez, G., and Normant, J., *J. Organomet. Chem.*, 1979, **177**, 293, and references therein.

Marfat, A., McGuirk, P. R., Kramer, R., and Helquist, P., *J. Am. Chem. Soc.*, 1977, **99**, 253; *J. Org. Chem.*, 1979, **44**, 1345.

Marfat, A., McGuirk, P. R., and Helquist, P., *Tetrahedron Lett.*, **1978**, 1363.

Organocopper 1,6-Addition Reactions Since 1975

Davis, B. R. and Johnson, S. J., *J. Chem. Soc., Chem. Commun.*, **1978**, 614.

Organocopper 1,7-Addition to γ,δ-Vinyl-α,β-cyclopropyl Carbonyl Compounds Since 1975

Schultz, A. G., Godfrey, J. D., Arnold, E. V., and Clardy, J., *J. Am. Chem. Soc.*, 1979, **101**, 1276.

Taber, D. F., *J. Am. Chem. Soc.*, 1977, **99**, 3513.

Kondo, K., Umemoto, T., Takahatake, Y., and Tunemoto, D., *Tetrahedron Lett.*, **1977**, 113.

E. Acetylenic Carbonyl Compounds

Synthesis 9: Insect Juvenile Hormones

In contrast to 44,000, which is the total number of vertebrate animal species, approximately 800,000 species of insects exist on earth today. The major groups are beetles (280,000 species); wasps, bees, and ants (115,000); flies, gnats, and mosquitoes (87,000); and butterflies and moths (140,000). Because of their diversity and great numbers, insects contribute significantly to the ecological balance in nature. Insects, for example, are very important for cross-pollination and therefore reproduction of flowering plants. Some insects, however, are a threat to humans because they carry disease-causing organisms (e.g., malaria, sleeping sickness, and yellow fever) or because they destroy plants that are important to us. In 1978, for example, approximately 5 million spruce trees in Norway were killed by bark beetles; the economic value of the lost wood was about 200 million dollars. Likewise, the elm bark beetle is responsible for the spread of Dutch elm disease throughout the United States, and the Japanese beetle annually destroys millions of dollars worth of plants and vegetables in the U.S. Citrus fruits in the western part of the U.S. and in several Mediterranean countries are extensively damaged by Homopteran insects, and in 1970 alone 200 million dollars in damage to the U.S. cotton crop was caused by the boll weevil. Insect pest control, therefore, is of great practical and economic importance to our modern society, and the agrochemical industry as well as the U.S. Department of Agriculture are vitally involved in this area.

Since the 1940s, many synthetic insecticides—organochlorines, organophosphates, and carbamates—have been and are continuing to be extraordinarily successful in controlling insects that imperil our supply of natural raw materials and foods. Because some insects have developed a degree of resistance to these pesticides and, more importantly, because many of these long-lasting insecticides are harmful to humans and animals, new and relatively safe ways of controlling insect pests are constantly being sought.

One powerful and effective new concept involves using insects' own natural growth-regulating hormones to control (i.e., delay or initiate) metamorphosis through the developmental stages of egg, larva, pupa, and finally adult. Since the middle 1960s, various insect growth-regulating hormones have been isolated, characterized, and eventually synthesized on a sufficiently large scale for field testing. The two most important hormones governing insect morphogenesis through a series of moults are ecdysone, which promotes growth and differentiation toward the adult, and juvenile hormone, which retards maturation. Clearly, intentional disruption of an insect's normal life cycle by dispersing man-made insect hormones that are not toxic to humans, animals, or plants is a potentially effective, practical, and specific means of insect pest control. Exposing an insect to more than its normal level of ecdysone during its larval stage, for example, speeds the insect's growth so much that it dies. In contrast, an overdose of juvenile hormone stops maturation at the larval stage or at least prolongs the larval stage so that immature and nonfunctional "adults" are eventually formed.

Another powerful and effective new concept involves using insects' natural communication hormones (i.e., pheromones) to control their behavior. We discuss insect pheromones in syntheses 12, 13, 18, 21–24, 27, 28, and 31.

Several different natural and synthetic juvenile hormones are known: 9-1a–9-1d. The major challenge in synthesizing these acyclic diolefinic epoxides is controlling the stereochemistry of the isolated trisubstituted double bonds; the terminal epoxide is formed easily from the corresponding triene. Although many new and useful methods have been developed to meet this challenge and several reviews are available, we focus attention in this book only on those methods involving organocopper reagents (cf. synthesis 10, 13, and 24). Specifically, stereocontrolled cis-addition of organocopper reagents across the triple bond of acetylenic esters was developed in 1968 as a key step in preparing juvenile hormones; retrosynthetic scheme 9-1 is therefore generated. As discussed in the original literature, the amount of stereoselectivity in the organocopper addition reaction depends on the reaction temperature and time, on the nature of the organocopper reagent, and on the solvent and ligands used. Polymeric RCu reagents prepared from Grignard reagents and copper(I) salts generally give the best results.

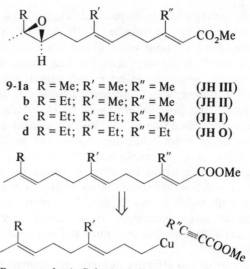

9-1a R = Me; R' = Me; R" = Me   (JH III)
   b R = Et; R' = Me; R" = Me   (JH II)
   c R = Et; R' = Et; R" = Me   (JH I)
   d R = Et; R' = Et; R" = Et   (JH O)

Retrosynthetic Scheme 9-1

Organocopper Reagents in the Synthesis of Trisubstituted Double Bonds via Addition to Acetylenic Esters

(a) Corey, E. J. and Katzenellenbogen, J. A., *J. Am. Chem. Soc.*, 1969, **91**, 1851.

(b) Siddall, J. B., Biskup, M., and Fried, J. H., *J. Am. Chem. Soc.*, 1969, **91**, 1853.

(c) Näf, F. and Degen, P., *Helv. Chim. Acta*, 1971, **54**, 1939.

(d) Corey, E. J., Kim, C. U., Chen, R. H. K., and Takeda, M., *J. Am. Chem. Soc.*, 1972, **94**, 4396.

(e) Anderson, R. J., Corbin, U. C., Cotterrell, G., Cox, G. R., Henrick, C. A., Schaub, F., and Siddall, J. B., *J. Am. Chem. Soc.*, 1972, **94**, 5374.

A Codling Moth Constituent

(f) Bowlus, S. J. and Katzenellenbogen, J. A., *Tetrahedron Lett.*, 1973, 1277.

Siccanin, an Antifungal Antibiotic

(g) Oida, S., Ohashi, Y., and Ohki, E., *Chem. Pharm. Bull.*, 1973, **21**, 528.

(2S, 3S)-[4-^{13}C] Valine and β-Lactam Antibiotics

(h) Kluender, H., Bradley, C. H., Sih, C. J., Fawcett, P., and Abraham, E. P., *J. Am. Chem. Soc.*, 1973, **95**, 6149.

A Proposed Codling Moth Pheromone

(i) Cooke, M. P., *Tetrahedron Lett.*, **1973**, 1281.

5,6-Dihydro-2H-pyran-2-ones

(j) Carlson, R. M., Oyler, A. R., and Peterson, J. R., *J. Org. Chem.*, 1975, **40**, 1610.

Farnesol

(k) Pitzele, B. S., Baran, J. S., and Steinman, D. H., *Tetrahedron*, 1976, **32**, 1347.

(l) Bryson, T. A., Smith, D. C., and Kruger, S. A., *Tetrahedron Lett.*, **1977**, 525.

α,γ-Diethylenic Esters

(m) Miginiac, P., Daviaud, G., and Gerard, F., *Tetrahedron Lett.*, **1979**, 1811.

Use of New RCu-BR$_3$ Reagents

(n) Yamamoto, Y., Yatagai, H., and Maruyama, K., *J. Org. Chem.*, 1979, **44**, 1745.

(o) Miginiac, P., Daviaud, G., and Gérard, F., *Tetrahedron Lett.,* **1979**, 1811.

(±)-Heliotridine and (±)-Retronecine

(p) Keck, G. and Nickell, D. G., *J. Am. Chem. Soc.,* 1980, **102**, 3632.

General Reading

Gilbert, L. I., Ed., *The Juvenile Hormones*, Plenum Press, New York, 1976.

Stern, E. S., Cavalla, J. F., and Price-Jones, D., *The Chemist in Industry (2): Human Health and Plant Protection*, Clarendon Press, Oxford, England, 1974.

Novak, V. J. A., *Insect Hormones*, Methuen & Co., Ltd., London, England, 1966.

Karlson, P., *Pure Appl. Chem.*, 1967, **14**, 75.

Reviews of Stereoselective Synthesis of Trisubstituted Double Bonds:

(a) Faulkner, D. J., *Synthesis*, **1971**, 175.

(b) Reucroft, J. and Sammes, P., *Q. Rev.*, **1971**, 135.

F. Acetylenes

Synthesis 10: Insect Juvenile Hormones

A second organocopper approach to synthesis of juvenile hormones was developed independently by French, Dutch, and American chemists. This approach involves addition of organocopper reagents to isolated (i.e., noncarbonyl conjugated) acetylenes and subsequent carboxylation or alkylation of the intermediate vinylcopper species. The overall two-step sequence amounts to stereocontrolled *cis*-addition of a nucleophilic group R and an electrophilic group across a carbon-carbon triple bond with formation of a trisubstituted olefin, eq. 10-1.

$$-C{\equiv}CH \xrightarrow{\ RCu\,\cdot\,SMe_2\ } \underset{R}{\overset{H}{\diagup\!\!\diagdown}}\underset{Cu}{} \xrightarrow{\ E^+\ } \underset{R}{\overset{H}{\diagup\!\!\diagdown}}\underset{E}{} \qquad (10\text{-}1)$$

Iterative retrosynthetic scheme 10-1 is based on this approach. The $RCu\cdot SMe_2$ reagent is formed from the $Me_2S\cdot CuBr$ complex and RMgX. The $Me_2S\cdot CuBr$ complex can be purfied by crystallization, which eliminates impurities causing decomposition of the intermediate vinylic copper species.

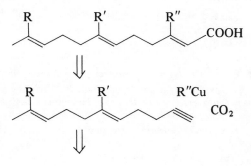

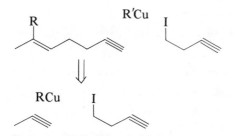

Retrosynthetic Scheme 10-1

References for Further Reading

French Contributions

Normant, J. F., *Pure Appl. Chem.*, 1978, **50**, 709.

Normant, J. F., *J. Organomet. Chem. Libr.*, 1976, **1**, 219.

Chuit, C., Cahiez, G., and Normant, J. F., *Tetrahedron*, 1976, **32**, 1675.

Alexakis, A., Cahiez, G., and Normant, J. F., *J. Organomet. Chem.*, 1979, **177**, 293.

Alexakis, A., Cahiez, G., and Normant, J. F., *Synthesis*, 1979, 826.

Insect Pheromone Bombykol

Normant, J. F., Commercon, A., and Villiéras, J., *Tetrahedron Lett.*, **1975**, 1465.

Normant, J. F. and Bourgain, M., *Tetrahedron Lett.*, **1971**, 2583.

Dutch Contributions

Westmijze, H., Meijer, J., Bos, H. J. T., and Vermeer, P., *Rec. Trav. Chim.*, 1976, **95**, 299, 307.

Westmijze, H., Kleijn, H., and Vermeer, P., *Tetrahedron Lett.*, **1977**, 2023.

Westmijze, H., Kleijn, H., and Vermeer, P., *Synthesis*, 1978, 454.

American Contributions

Marfat, A., McGuirk, P. R., Kramer, R., and Helquist, P., *J. Am. Chem. Soc.*, 1977, **99**, 253.

Lalima, N. J. and Levy, A. B., *J. Org. Chem.*, 1978, **43**, 1279.

Marfat, A., McGuirk, P. R., and Helquist, P., *J. Org. Chem.*, 1979, **44**, 1345, 3888.

Japanese Contribution

  Obayashi, M., Utimoto, K., and Nozaki, H., *J. Organomet. Chem.*, 1979,
  177, 145.

Application to Synthesis of an Oxido-Eicosopentanoic Acid

  Corey, E. J., Arai, Y., and Mioskowski, C., *J. Am. Chem. Soc.*, 1979, **101**,
  6748.

G. Miscellaneous

Synthesis 11: Geraniol, A Perfume Constituent

Geraniol **(11-1)** is one of the most frequently used perfume chemicals;
it is used in delicate lotion perfumes and soft cosmetic fragrances as
well as in sweet floral household scents and inexpensive soap perfumes.
In very low concentrations geraniol also is used in industrial fabrication
of many different food flavor types, including apple, apricot, straw-
berry, raspberry, plum, peach, honey, cherry, lemon, cassis, cinnamon,
nutmeg, root beer, and ginger ale. Geraniol is found in nature as the
main constituent of rose oil and palmarosa oil and as a minor constituent of other essential oils such as citronella and lemon grass.

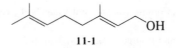

11-1

Being a 10-carbon diene alcohol, geraniol can be synthesized in a
large number of different and effective ways. In this book we focus
attention only on those reported synthetic approaches that use organocopper reagents. In Synthesis 17 we describe coupling of two allylic
groups to produce the 1,5-diene unit of geraniol, but here we consider
retrosynthetic scheme **11-1**, involving manipulation of the allylic alcohol
portion of geraniol.

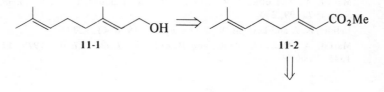

11-1   **11-2**

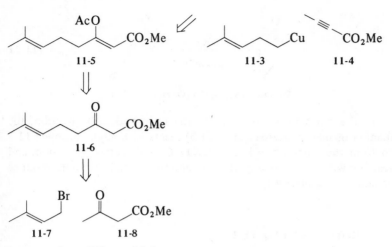

Retrosynthetic Scheme 11-1

The basic simplification in retrosynthetic scheme 11-1 involves considering the allylic alcohol functionality as synthetically equivalent to (i.e., preparable from) the corresponding ethylenic ester **11-2**. Conjugated ethylenic ester **11-2** now can be analyzed in terms of an organocopper conjugate *syn*-addition reaction to acetylenic ester **11-4** (*cf.* syntheses 9 and 10: juvenile hormones) or in terms of an overall replacement of acetoxy by methyl in β-acetoxy α,β-ethylenic ester **11-5** (i.e., eq. 11-1). Conversion of β-acetoxy enoate ester **11-5** into β-methyl enoate ester **11-2** might be considered at first glance to be a direct substitution reaction, but it is more probably a conjugate addition-elimination sequence.

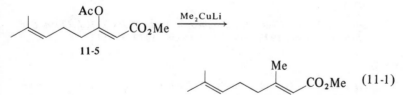

Many examples have appeared since 1973 illustrating the utility of organocopper reagents in replacing β-$\ddot{Y}$ groups in α,β-ethylenic carbonyl compounds, as shown in eq. 11-2.

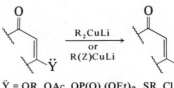

$$\ddot{Y} = OR, OAc, OP(O) (OEt)_2, SR, Cl, I$$

$$(11\text{-}2)$$

In the synthetic (as opposed to the retrosynthetic) direction, the dianion of methyl acetoacetate (11-8) reacts with allylic bromide 11-7 to form the nine-carbon keto ester 11-6; enol acetate formation, and reaction with dimethylcopperlithium and then with LiAlH$_3$OEt produce geraniol stereoselectively.

References for Eq. 11-2

Y = OR

    (a) Posner, G. H. and Brunelle, D. J., *J. Chem. Soc., Chem. Commun.,* **1973**, 907.

    (b) Cacchi, S., Caputo, A., and Misiti, D., *Indian J. Chem.,* **1974**, 325.

Y = OAc

    (a) Casey, C. P., Marten, D. F., and Boggs, R. A., *Tetrahedron Lett.,* **1973**, 2071.

    (b) Casey, C. P. and Marten, D. F., *Synth. Commun.,* **1973**, 3, 321.

    (c) Ouannes, C. and Langlois, Y., *Tetrahedron Lett.,* **1975**, 3461.

$$\overset{O}{\underset{\|}{}}$$
Y = OP(OEt)$_2$

    Sum, F. W. and Weiler, L., *Tetrahedron Lett.,* **1979**, 707; *J. Chem. Soc., Chem. Commun.,* **1978**, 985; *Can. J. Chem.,* **1979**, 57, 1431.

Y = SR

    (a) Posner, G. H. and Brunelle, D. J., *J. Chem. Soc., Chem. Commun.,* **1973**, 907.

    (b) Kobayashi, S. and Mukaiyama, T., *Chem. Lett.,* **1974**, 705; **1973**, 1097.

    (c) Corey, E. J. and Chen, H. K., *Tetrahedron Lett.,* **1973**, 3817.

$Y = Cl$

(a) Clark, R. D. and Heathcock, C. H., *J. Org. Chem.,* 1976, **41,** 636.

(b) Coke, J. L., Williams, H. J., and Natarajan, S., *J. Org. Chem.,* 1977, **42,** 2380.

(c) Wender, P. A. and Eck, S. L., *Tetrahedron Lett.,* 1977, 1245.

(d) Leyendecker, F., Drouin, J., and Conia, J. M., *Tetrahedron Lett.,* 1974, 2931.

$Y = I$

Piers, E. and Morton, H. E., *J. Org. Chem.,* 1979, **44,** 3437; *J. Chem. Soc., Chem. Commun.,* 1978, 1033.

3

Organocopper Substitution Reactions

A. Alkyl Halides

Synthesis 12: Z-9-Tricosene (Muscalure), A Housefly Sex Attractant

Animals and insects use chemicals called *pheromones* to alter the behavior of other animals and insects. We need only mention the skunk to remind us that humans also are subject to the effects of some pheromones. Many pheromones, however, are species specific, affecting only or mainly members of the same species. There are various types of pheromones, including food pheromones, trail pheromones, aggregating pheromones, alarm pheromones, recruiting pheromones, and sex pheromones. As an example of a recruiting pheromone, a chemical odor sectreted by termite or ant larvae attracts workers who then, by instinct, care for and protect the larvae. Insect sex attractant pheromones are often species-specific reproducible mixtures of several organic compounds, usually with one compound being the major component; insects normally produce and use these chemical sex attractants to help males and females locate one another and reproduce. Many successful field tests have been carried out using synthetic insect sex attractants to lure male insects into traps or to confuse (i.e., overload their guidance system) male insects so that they cannot locate any female insects; reproduction is therefore prevented.

By far the largest use of synthetic insect sex attractants took place in Norway in 1979, when 220 pounds of the Ips bark beetle sex pheromone were synthesized and distributed in 600,000 traps to protect the spruce trees there from beetle damage. The project, which cost 18 million dollars, represents a modern, sophisticated approach to management of insect pests using behavior-controlling chemicals.

The sexually mature femal housefly, *Musca domestica,* secretes a very simple hydrocarbon, Z-9-tricosene, which attracts sexually mature male houseflies; in field tests Z-9-tricosene (muscalure, **12-1**) also acts as an aggregating pheromone by attracting both male and female houseflies. This olefinic hydrocarbon is currently produced by the Zoëcon Corporation in Palo Alto, California, and it is used to increase the effectiveness of a fly bait containing insecticide; this fly bait (Golden Malrin^R) is the first commercial product using a sex attractant pheromone for insect control purposes which has been registered by the U.S. Environmental Protection Agency.

$$n\text{-}C_8H_{17} \quad C_{13}H_{27}\text{-}n$$

12-1

What method does Zoëcon use to prepare Z-9-tricosene on a commercial scale? Because Z-9-tricosene is a relatively large but uncomplicated organic compound, we can design a great number of synthetic approaches for preparing olefin **12-1**. Two reasonable approaches involve formation of the central *cis* double bond: (1) stereoselective Wittig reaction of an alkylidene phosphorane with the appropriate straight-chain aldehyde ($Z:E$ double bond is formed in $85:15$ ratio), and (2) acetylide coupling with a straight-chain alkyl bromide or iodide followed by triple bond *cis*-hydrogenation with a Lindlar catalyst to form the *cis* double bond predominantly.

Two other rational approaches start with naturally occurring, commercially avaliable long-chain 18-carbon or 22-carbon Z-olefinic carboxylic acids. These fatty acids are converted into the corresponding 23-carbon ketones, which are then reduced to Z-9-tricosene (eq. 21-1).

Because of the commercial availability of oleyl alcohol (and therefore the availability of oleyl bromide) and because of the general effectiveness of organocopper (in contrast to RMgBr or RLi) reagents in substitution reactions with alkyl halides, Zoëcon has developed the

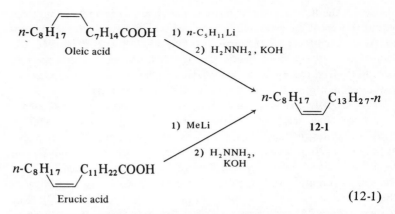

(12-1)

production-scale (200 pound batches) synthesis of muscalure (12-1) shown in eq. 12-2.

$$n\text{-}C_8H_{17} \quad C_7H_{14}CH_2Br \xrightarrow[\substack{0.03 \text{ equiv.,} \\ \text{Cl(CN)CuLi,} \\ \text{THF,} \\ 0-5°}]{\substack{1.15 \text{ equiv.,} \\ n\text{-}C_5H_{11}MgBr}} \text{12-1} \quad (12\text{-}2)$$

(99%)

Undoubtedly reaction 12-2 proceeds through the intermediacy of small amounts of hetero n-amyl(cyano)cuprate species which are stable between 0° and 5° in tetrahydrofuran solvent and yet reactive enough to couple with oleyl bromide and to generate a bromo(cyano)cuprate which then reacts with amyl Grignard and continues the catalytic cycle. The chloro(cyano)copperlithium catalyst was designed by Zoëcon to produce an amyl(*cyano*)cuprate which is more thermally stable than the corresponding amyl(*chloro*)cuprate formed from dichlorocopper-lithium; the 15° or so greater thermal stability of the amyl(*cyano*)-cuprate over the amyl(*chloro*)cuprate intermediate overcame a number of pragmatic problems dealing with maintenance of adequate cooling and heat transfer for plant-scale reactions run at ⁻10 to ⁻20°. We have mentioned the usefulness of heterocuprates before in the discussion of prostaglandins (synthesis 5).

Reaction 12-2 can also be achieved using the homocuprate (amyl)$_2$-CuLi or the corresponding mixed homocuprates amyl(R′)CuLi. Reaction 12-2, however, is carried out industrially via a copper-catalyzed coupling process because such a catalytic process allows higher operating temperatures and use of amyl Grignard rather than less stable amyllithium; this catalytic process is therefore cheaper.

References for Further Reading

RCOOH + R'Li ⟶ RCOR'

Jorgenson, M. J., *Org. React.,* 1970, **18**, 1.

Zoëcon's Preparation of Z-9-Tricosene and General Syntheses of Insect Sex Pheromones

Henrick, C. A., *Tetrahedron,* 1977, **33**, 1.

General Reading

Beroza, M., ed., *Pest Management with Insect Sex Attractants,* ACS Symposium Series, No. 23, American Chemical Society, Washington, D.C., 1976.

Marini-Beltolo, G. B., Ed., *Natural Products and the Protection of Plants: Proceedings of a Study Week of the Pontifical Academy of Sciences,* October, 1976, Elsevier Publishing Co., Amsterdam, 1978.

Rossi, R., "Insect Pheromones: Synthesis of Chiral Components of Insect Pheromones," *Synthesis,* 1978, 413.

Examples of Natural Product Synthesis Using Organocopper Coupling with Primary Alkyl Halides

A Codling Moth Sex Attractant

Descoins, C. and Henrick, C. A., *Tetrahedron Lett.,* 1972, 2999.

A European Grapevine Moth Sex Attractant

Labovitz, J., Henrick, C. A., and Corbin, V. L., *Tetrahedron Lett.,* 1975, 4209.

Fatty Acids

(a) Bergbreiter, D. E. and Whitesides, G. M., *J. Org. Chem.,* 1975, **40**, 779.

(b) Baer, T. and Carney, R. L., *Tetrahedron Lett.,* 1976, 4697.

German Cockroach Sex Pheromone

>   Rosenblum, L. D., Anderson, R. J., and Henrick, C. A., *Tetrahedron Lett.,*
>   **1976**, 419.

Red Bollworm Moth Sex Pheromone

>   Mandai, T., Yasuda, H., Kaito, M., and Tsuji, J., *Tetrahedron,* 1979, **35,**
>   309.

Bombykol Sex Pheromone

>   Samain, D. and Descoins, C., *Bull. Soc. Chim. Fr.,* **1979,** II-71.

Dendrolasin

>   Kojima, Y., Wakita, S., and Kato, N., *Tetrahedron Lett.,* **1979,** 4577.

Pine Sawflies Sex Attractant

>   Baker, R., Winton, P. M., and Turner, R. W., *Tetrahedron Lett.,* 1980,
>   **21,** 1175.

B. Alkenyl Halides

Synthesis 13: Insect Juvenile Hormones

We have already noted in syntheses 9 and 10 that juvenile hormones
can be prepared via organocopper additions to acetylenic esters and to
isolated acetylenes. The trisubstituted double bonds of juvenile hormones have also been formed from propargylic alcohols via organo-
aluminum and organocopper intermediates, as shown in eq. 13-1.

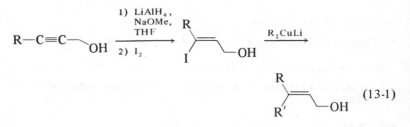

$$R-C{\equiv}C-OH \xrightarrow[\text{2) } I_2]{\substack{\text{1) } LiAlH_4, \\ NaOMe, \\ THF}} \quad \underset{I}{\overset{R}{>}}{=}-OH \xrightarrow{R_2CuLi}$$

$$\underset{R'}{\overset{R}{>}}{=}-OH \qquad (13\text{-}1)$$

The organocuprate coupling reaction shown in eq. 13-1 occurs with
complete retention of double bond geometry and in the presence of the

unprotected allylic hydroxyl group. In the original synthesis of juvenile hormone **13-2** using this approach, both the Et and Me groups shown in structure **13-1** were introduced separately via diorganocopperlithium substitution on the corresponding iodides.

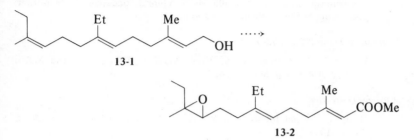

13-1

13-2

This ability of organocopper reagents to effect nucleophilic substitution of *alkenyl* halides is particularly impressive when compared with the classical inertness of alkenyl halides toward nucleophiles. Furthermore, the high stereochemical control exhibited in organocopper coupling reactions with alkenyl halides has been used as the key step in construction of many larger alkenes from smaller alkenyl halides.

Juvenile Hormone Synthesis

Corey, E. J., Katzenellenbogen, J. A., Gilman, N. W., Roman, S. A., and Erickson, B. W., *J. Am. Chem. Soc.,* 1968, **90**, 5618.

trans,trans-Farnesol

Corey, E. J., Katzenellenbogen, J. A., and Posner, G. H., *J. Am. Chem. Soc.,* 1967, **89**, 4245.

Organocopper Coupling with Alkenyl Halides to Form Natural Products

A Codling Moth Potential Pheromone

Bowlus, S. B. and Katzenellenbogen, J. A., *J. Org. Chem.,* 1973, **38**, 2733.
Cooke, M. P. Jr., *Tetrahedron Lett.,* **1973**, 1983.

Fulvoplumierin, an Antibacterial Metabolite

>   Buchi, G. and Carlson, J. A., *J. Am. Chem. Soc.,* 1969, **91**, 6420.
>
>   (This synthesis involves retention of configuration in the nucleophilic vinyl group transferred from the organocopper reagent and retention of configuration in the electrophilic vinylic chloride undergoing substitution; organocopper chemoselectivity is also illustrated.)

Shionine, a Tetracyclic Triterpene

>   Ireland, R. E., Lipinski, C. A., Kowalski, C. J., Tilley, J. W., and Walba, D. M., *J. Am. Chem. Soc.,* 1974, **96**, 3333.

A Corrin Derivative

>   Hamilton, A. and Johnson, A. W., *Chem. Commun.,* 1971, 323; *J. Chem. Soc. C,* 1971, 3879.

γ-Elemene

>   Kato, M., Kurihara, H., and Yoshikoshi, A., *J. Chem. Soc. Perkin I,* 1979, 2740.

Organocopper Coupling with 1-Bromo-1-trimethylsilylalkenes

Miller, R. B., and McGarvey, G., *J. Org. Chem.,* 1979, **44**, 4623.

C. Alkynyl Halides

Synthesis 14:   Freelingyne, An Acetylenic Sesquiterpene

Freelingyne **(14-1)** was the first natural acetylenic terpenoid to be isolated (1966). Since then a variety of natural acetylenic terpenoids have been isolated and identified. Because freelingyne is a very highly functionalized sesquiterpene, we can devise many reasonable approaches to its synthesis. We focus here, however, on an organocopper approach reported in 1974–1975 involving convergent joining of a four-carbon furyl group with a six-carbon acetylenic group and then with a five-carbon Wittig reagent (retrosynthetic scheme 14-1).

Connecting the furyl and acetylenic groups can be achieved in two complementary ways: (1) using a nucleophilic furyl group and an electrophilic acetylenic group (eq. 14-1), or (2) using an electrophilic

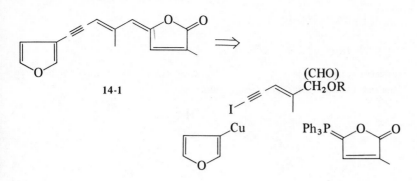

Retrosynthetic Scheme 14-1

furyl group and a nucleophilic acetylenic group (eq. 14-2). The most important and dramatic difference between these two reaction types is that reaction 14-1 proceeds at temperatures from $-45°$ to $+25°$, whereas reaction 14-2 requires hot dimethylformamide ($120°$) or refluxing pyridine ($114°$) to proceed effectively.

Furthermore, when the aryl group Ar is 3-furyl, eq. 14-2 provided substantial amounts of dimeric bisfurans and diynes along with the desired furyl-acetylenic coupling product.

$$Ar-Cu + IC{\equiv}CR \xrightarrow[\substack{-45° \\ \text{to} \\ +25°}]{} Ar-C{\equiv}CR \qquad (14\text{-}1)$$

$$Ar-I + CuC{\equiv}CR \xrightarrow[\substack{DMF, \\ reflux}]{} Ar-C{\equiv}CR \qquad (14\text{-}2)$$

Organocopper Synthesis of Freelingyne

Knight, D. W. and Pattenden, G., *J. Chem. Soc., Perkin I*, 1975, 641.

Examples of Arylcopper Coupling with Alkynyl Iodides

(a) Oliver, R. and Walton, D. R. M., *Tetrahedron Lett.*, 1972, 5209.

(b) Verboom, W., Westmijze, H., Bos, H. J. T., and Vermeer, P., *Tetrahedron Lett.*, 1978, 1441.

D. Aryl Halides

Synthesis 15: Frustulosin, An Antibiotic

Frustulosin **(15-1)** is an antibiotic that shows antimicrobial activity against a broad range of pathogenic bacteria.

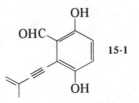

Like the arylacetylene freelingyne discussed in synthesis 14, aryl-acetylene frustulosin can be assembled most directly via coupling of an aryl group and an alkynyl group. The presence of a reactive aldehyde functionality dictates the choice of an aryl iodide and a cuprous acetyl-ide rather than an arylcopper species and an iodoacetylene; arylcopper species add across aldehyde carbonyl groups, but cuprous acetylides do not do so even under forcing conditions. Cuprous isopropenylacetylide couples with *o*-formyliodobenzene **15-2** in refluxing DMF to give frustulosin directly (eq. 15-1). Indeed, many different aryl and hetero-aryl acetylenes have been assembled via coupling of the corresponding cuprous acetylides and aryl or heteroaryl halides.

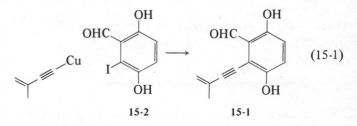

(15-1)

Natural Products Formed via Coupling of Cuprous Acetylides with Aryl Halides

Junipal

Atkinson, R. E., Curtis, R. F., and Phillips, F. T., *Chem. Ind.*, **1964**, 2101.

Frustulosin

>   Ronald, R. C. and Lansinger, J. M., *J. Chem. Soc., Chem. Commun.*, 1979, 124.

Rotenic acid

>   Batu, G. and Stevenson, R., *J. Org. Chem.*, 1979, 44, 2948.

Artemidin

>   Batu, G. and Stevenson, R., *J. Org. Chem.*, 1980, 45, 1532.

Key References to Cuprous Acetylide Preparation and Coupling with Aryl Halides

>   (a) Posner, G. H., *Org. React.*, 1975, 22, 253.
>   (b) Castro, C. E., Havlin, R., Honwad, V. K., Malte, A., and Mojé, S., *J. Am. Chem. Soc.*, 1969, 91, 6464.
>   (c) Castro, C. E., Gaughan, E. J., and Owsley, D. C., *J. Org. Chem.*, 1966, 31, 4071.

Some Examples of Organocopper Coupling with Aryl Halides Since 1975

>   (a) King, F. D. and Walton, D. R. M., *Synthesis*, 1976, 40.
>   (b) Kraus, G. and Frazier, K., *Tetrahedron Lett.*, 1978, 3195.
>   (c) Duffley, R. P. and Stevenson, R., *Synth. Commun.*, 1978, 8, 175.
>   (d) Schreiber, F. G. and Stevenson, R., *Org. Prep. Proc. Int.*, 1978, 10, 137.
>   (e) Cornforth, J., Sierakowski, A., and Wallace, T. W., *J. Chem. Soc., Chem. Commun.*, 1979, 294.
>   (f) Ziegler, F. E., Chliwner, I., Fowler, K. W., Kamfer, S. J., Kuo, S. J., and Sinha, N. D., *J. Am. Chem. Soc.*, 1980, 102, 790.

E. Benzylic and Allylic Halides

Synthesis 16:   Olivetol, An Intermediate in the Synthesis of Hashish

Olivetol (16-1), a phenol isolated from lichens, showed some promise in the 1940s of being a useful antibiotic against pneumococci and diphtheria bacteria as well as a toxic agent for leeches. Little use was made of olivetol, however, until the late 1960s, when it was recognized that tetrahydrocannabinols 16-3 (the most active hashish components), can be prepared by joining symmetrical olivetol with *p*-menthadienol **16-2** under acidic conditions (retrosynthetic scheme 16-1).

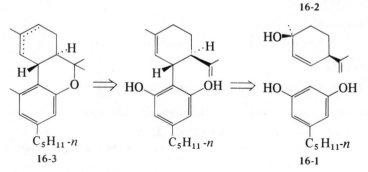

Retrosynthetic Scheme 16-1

Because benzylic halides are normally very reactive toward nucleophilic displacement, olivetol can be analyzed as shown in retrosynthetic scheme 16-2. Benzylic bromide **16-4** is readily available from natural resorcylic acid via eq. 16-1. Although the dibutylcopperlithium coupling was actually performed on the bismethoxy derivative **16-4**, this substitution reaction would probably work also in the presence of unprotected phenol hydroxyl groups. As with most organocopper coupling reactions with benzylic halides, olivetol is accompanied by some bibenzyl side-product, presumably arising from a benzylcopper intermediate.

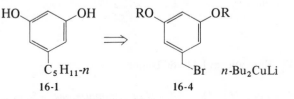

Retrosynthetic Scheme 16-2

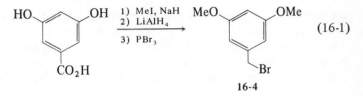

Organocopper Route to Olivetol

Also Olivetol → Tetrahydrocannabinols

Petrzilka, T., Haefliger, W., and Sikemeier, C., *Helv. Chim. Acta*, 1969, **52**, 1102.

References for Further Reading

Islandicin Methyl Ether (Benzylic Coupling)

Whitlock, B. J. and Whitlock, H. W., *J. Org. Chem.*, 1980, **45**, 12.

Synthesis 17:   Geraniol, A Perfume Constituent

Use of geraniol as a perfume constituent and as a food flavor was described in synthesis 11. Retrosynthetic analysis (scheme 17-1) of 10-carbon geraniol leads directly to a pair of five-carbon units, one an allylic electrophile and the other an allylic nucleophile.

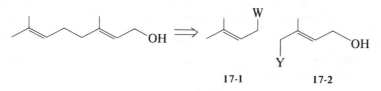

$$\text{17-1} \qquad\qquad \text{17-2}$$

W = metal, and Y = halogen or sulfonate
W = haolgen or sulfonate, and Y = Metal

Retrosynthetic Scheme 17-1

Of the two possible polarities for five-carbon units **17-1** and **17-2**, successful construction of geraniol has been reported only for nucleophilic **17-1** and electrophilic **17-2**. When **17-1** is a Grignard reagent (W = MgCl), an equilibrium is set up between two isomeric Grignard reagents, and reaction with alkyl tosylates yields mainly terminal olefins (eq. 17-1). In contrast, when Grignard reagent **17-1**, W = MgCl is treated with 0.1 equiv. of cuprous iodide, coupling with alkyl electrophiles occurs rapidly and exclusively at the primary carbon atom (e.g., eq. 17-2). Equation 17-2 represents a convergent and highly efficient synthesis of terpene geraniol.

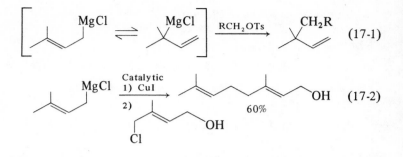

References

Geraniol

   Derguini-Boumechal, F., Lorne, R., and Linstrumele, G., *Tetrahedron Lett.*, **1977**, 1181.

Potato Tubeworm Moth Sex Pheromone (Cuprous Acetylide Coupling with an Allylic Chloride)

   Roelofs, W. L., Kochansky, J. P., Carde, R. T., Kennedy, G. G., Henrick, C. A., Labovitz, J. N., and Corbin, V. L., *Life Sci.*, **1975**, **17**, 699.

Cymopol

   Raynolds, P. W., Manning, M. J., and Swenton, J. S., *J. Chem. Soc., Chem. Commun.*, **1977**, 499.

Menaquinones

   Snyder, C. D. and Rapoport, H., *J. Am. Chem. Soc.*, **1974**, **96**, 8046.

References to Organocopper Coupling with Benzylic and Allylic Halides Since 1975.

(a) Grieco, P. A., Wang, C-L. J., and Majetich, G., *J. Org. Chem.*, **1976**, **41**, 726.

(b) Yamamoto, Y., Yatagai, H., Sonoda, A., and Murahashi, S., *J. Chem. Soc., Chem. Commun.*, **1976**, 452.

(c) Ziegler, F. E. and Fowler, K. W., *J. Org. Chem.*, **1976**, **41**, 1564.

(d) Gordon-Gray, C. G. and Whitely, C. G., *J. Chem. Soc., Perkin I*, **1977**, 2040.

(e) Itoh, K., Fukui, M., and Kurachi, Y., *J. Chem. Soc.*, 1977, 500.

(f) Savignac, P., Breque, A., Mathey, F., Varlet, J. M., and Collignon, N., *Synth. Commun.*, 1979, 9, 287.

F. Acyl Halides

Synthesis 18: Manicone, A Male Ant Pheromone

We have discussed pheromones in synthesis 12. Manicone [(*E*)-4,6-dimethyl-4-octen-3-one, **18-1**] is secreted from the mandibular gland of male ants of several species in the genus *Camponotus*. Manicone causes female ants to swarm at the same time as the males are swarming, and thus ant pairing, mating, and reproducing are coordinated by release of this pheromone.

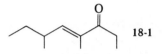

The 10-carbon skeleton of manicone can be analyzed in terms of a variety of reasonable and practical synthetic approaches; the α,β-ethylenic functionality offers many modes of access to this terpene. Perhaps the most obvious approach is shown in retrosynthetic scheme 18-1; in the synthetic direction this scheme involves an aldol condensation between the enolate ion (or the phosphonium ylide) of *symmetrical* 3-pentanone and a five-carbon aldehyde.

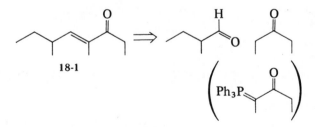

Retrosynthetic Scheme 18-1

One potential drawback to this approach is the likely formation of some of the unwanted *Z*-double bond geometrical isomer. As a more effective method for controlling double bond geometry, organocopper stereocontrolled *syn*-addition to an α,β-acetylenic acid or ester can be used (*cf.* synthesis 9); we are led therefore to retrosynthetic scheme 18-2. An acetylenic *ester* rather than an acetylenic *ketone* is chosen, because the methodology for organocopper *syn*-addition to acetylenic ketones is only now being perfected (in Japan).

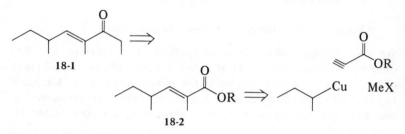

Retrosynthetic Scheme 18-2

In the synthetic direction, how can α,β-unsaturated acid **18-2** (R = H) be converted into the corresponding ethyl ketone (**18-1**)? One of the experimentally simplest and most useful applications of organocopper reagents has been in the transformation of acyl halides (chlorides) into ketones; the yields are high, many other functionalities (e.g., esters, nitriles, ketones, aryl halides, alkyl halides) are stable, the reaction is rapid and mild (−78°, 15 min), and the organocopper reagents are more easily prepared and less toxic than the classical organocadmium reagents. As illustrated in eq. 18-1, even an α,β-ethylenic ketone can be prepared in this way, especially when the ketone is α,β-disubstituted so that organocopper conjugate addition is severely retarded.

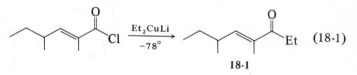

References

Manicone Isolation

    Brand, J. M., Duffield, R. M., MacConnell, J. G., Blum, M. S., and Fales, H. M., *Science,* 1973, *179,* 388.

Organocopper Syn-Addition to Proprolic Acid and Esters and to Acetylenic Ketones

Yamamoto, Y., Yatagai, H., and Maruyama, K., *J. Org. Chem.,* 1979, **44**, 1745.

Eq. 18-1

. Katzenellenbogen, J. A. and Utawonit, T., *J. Am. Chem. Soc.,* 1974, **96**, 6153; Mimura, T., Kimura, Y., and Nakai, T., *Chem. Lett.,* 1979, 1361.

Organocopper Conversion of Acyl Halides into Ketones

(a) Posner, G. H., Whitten, C. E., and McFarland, P. E., *J. Am. Chem. Soc.,* 1972, **94**, 5106.

(b) Posner, G. H., Brunelle, D. J., and Sinoway, L., *Synthesis,* 1974, 662.

(c) Posner, G. H. and Whitten, C. E., *Org. Synth.,* 1976, **55**, 122.

(d) Miyaura, N., Susaki, N., Itoh, M., and Susuki, A., *Tetrahedron Lett.,* 1977, 173.

(e) Piers, E. and Reissig, U., *Ang. Chem. Int. Ed. Eng.,* 1979, **18**, 791.

Cubanes
Dauben, W. G., and Reitman, L. N., *J. Org. Chem.,* 1975, **40**, 835.

Velleral Skeleton
Fex, T., Foroborg, J., Magnuson, G., and Thoren, S., *J. Org. Chem.,* 1976, **41**, 3518.

Prostaglandins
Nakamura, N. and Sakai, K., *Tetrahedron Lett.,* 1978, 1549.

Cerulenin
Boeckman, R. K., Jr. and Thomas, E. W., *J. Am. Chem. Soc.,* 1979, **101**, 987.

Conversion of Acyl Halides into Ketones with Other Organo-transition Metal Reagents

Cd
Shirley, D. A., *Org. React.,* 1958, **8**, 28.

Rh
(a) Hegedus, L. S., Lo, S. M., and Bloss, E. E., *J. Am. Chem. Soc.,* 1973, **95**, 3040.

(b) Pittman, C. U., Jr. and Hanes, R. M., *J. Org. Chem.,* 1977, **42**, 1194.

Pd
(a) Takagi, K. and Okamoto, T., *Chem. Lett.,* 1975, **9**, 951.

(b) Milstein, D. and Stille, J. K., *J. Am. Chem. Soc.,* 1978, **100**, 3636.

Fe
(a) Sawa, Y., Ryang, M., and Tsutsumi, G., *J. Org. Chem.,* 1970, **35**, 4183.
(b) Collman, J. P. and Hoffman, N. W., *J. Am. Chem. Soc.,* 1973, **95**, 2689.

Mn
Cahiez, G., Bernard, D., and Normant, J. F., *Synthesis,* 1977, 130.

Synthesis 19:   Fusidic Acid, A Steroidal Antibiotic

Fusidic acid (**19-1**), a relatively new structural type of tetracyclic triterpene, is active against infections caused by staphylococci. Although fusidic acid itself has not been prepared using organocopper reagents, the fusidic acid tetracyclic ring system **19-2** has; tetracycle **19-2** is actually a degradation product of fusidic acid and can also serve as an intermediate in a total synthesis of the antibiotic.

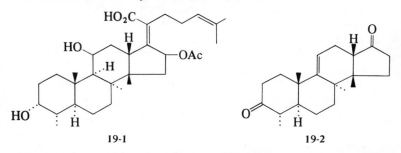

One of the key steps in the reported synthesis of tetracycle **19-2** involves annelating a fourth ring onto a tricycle. The method chosen for this annelation requires formation of a 1,5-diketone and subsequent intramolecular aldol condensation to form the fourth ring. The 1,5-diketone is produced via organocopper high-yield coupling *chemospecifically* with the acyl chloride portion of keto diene ether **19-3** (eq. 19-1).

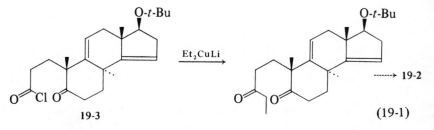

Construction of Steroidal Tetracycle 19-2

Dauben, W. G., Ahlgren, G., Leitereg, T. J., Schwarzel, W. C., and Yoshioko, M., *J. Am. Chem. Soc.*, 1972, **94**, 8593.

Construction of Steroidal Glycyrrhetic Ketones from Acyl Chlorides and Organocopper Reagents

Rozen, S., Shahak, I., and Bergmann, E. D., *Synthesis*, 1972, 701.

(The organocopper method works well, whereas organolithium reaction with the corresponding steroidal carboxylic acid or its lithium salt fails, as does organocadmium reaction with the acid chloride.)

G. Alkyl Sulfonate Esters

Synthesis 20: Vitamin E (α-Tocopherol)

Vitamin E (α-tocopherol, **20-1**, Gr. *tokos,* childbirth, + Gr. *pherein,* to bear) is one of the vitamins essential for normal human growth and development. It is involved especially in regulating the reproductive system and is sometimes used therapeutically in certain phases of infertility. The U.S. recommended daily allowance (RDA) of Vitamin E is 10-30 mg, and it is commonly added to a variety of foods even though it occurs naturally, especially in wheat germ, alfalfa, and lettuce. The tocopherols are thought to function in animals also as antioxidants for fats.

Pure α-tocopherol was isolated in 1936 and was first synthesized in the laboratory in 1938. Since then, many successful synthetic approaches to racemic α-tocopherol have been developed, with the first total synthesis of natural, optically active α-tocopherol being reported in 1963. As U.S. government regulations of food additives become stricter, the need increases for synthesis of large quantities of the biologically active enantiomers of various substances important for good health. A highly efficient and conceptually elegant approach to optically active α-tocopherol has been developed by chemists at Hoffmann-La Roche; this approach, represented in the retrosynthetic direction in scheme 20-1, illustrates the growing use of microbiological reactions in organic synthesis and also the effectiveness of copper catalysis in

joining two different carbon skeletons by forming a carbon-carbon
σ-bond between them. Specifically, scheme 20-1 relies on copper-
catalyzed coupling of Grignard reagents with alkyl *p*-toluenesulfonate
esters, a reaction involving transient organocopper intermediates, pop-
ularized by Fouquet and Schlosser in 1974.

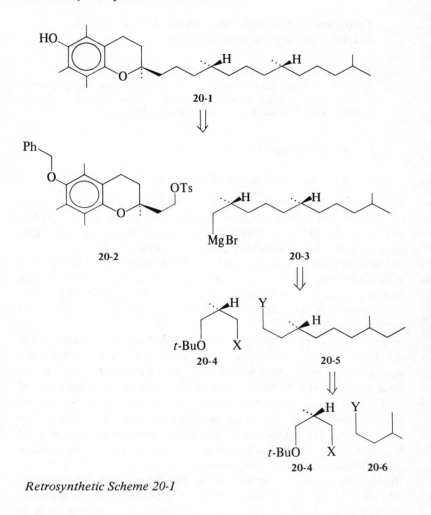

Retrosynthetic Scheme 20-1

Copper-catalyzed coupling of optically active 14-carbon Grignard reagent **20-3** with chroman *p*-tosylate **20-2** produced the α-tocopherol skeleton in 93% yield. How was the chiral 14-carbon Grignard reagent **20-3** prepared?

First, using a regiospecific and stereospecific microbiological oxidation of isobutyric acid to produce optically active 4-carbon synthon **20-4** (eq. 20-1), the Hoffmann-La Roche chemists then used synthon **20-4** either as the corresponding *p*-tosylate to couple with 10-carbon Grignard reagent **20-5** (Y = MgBr) or as the corresponding 4-carbon Grignard reagent to couple with 10-carbon *p*-tosylate **20-5** (Y = OTs, cf. eq. 20-2). It is worth noting that eq. 20-2, used to form the key 14-carbon intermediate, proceeds equally well from the left as from the right; umpolung (charge-affinity inversion) therefore has no practical advantage in this case.

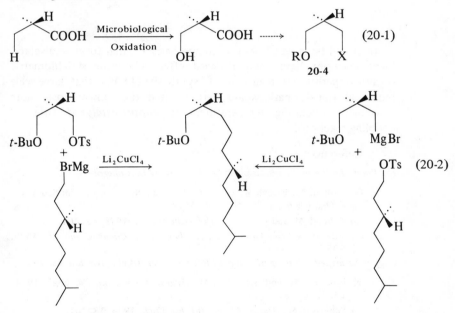

Further efficient use was made of optically active bifunctional 4-carbon synthon **20-4** by joining it with a 5-carbon and then with a

1-carbon unit to prepare the required optically active 10-carbon synthon **20-5** (eq. 20-3). In this way both of the chiral secondary methyl centers in α-tocopherol were derived from the *same* chiral 4-carbon synthon **20-4**.

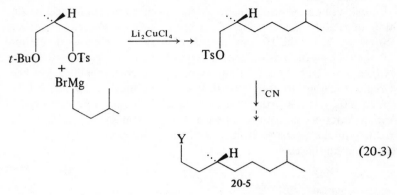

$$(20\text{-}3)$$

20-5

It should be stressed here, as in synthesis 12, that tosylate displacement reactions have been achieved effectively using stoichiometric organocopperlithium reagents (cf. synthesis 21) but that large-scale industrial transformations are better suited (i.e., lower cost, more convenient operating temperatures) to *copper-catalyzed* Grignard coupling reactions.

References

Hoffmann-La Roche Synthesis of Optically Active α-Tocopherol

(a) Cohen, N., Eichel, W. F., Lopresti, R. J., Neukom, C., and Saucy, G., *J. Org. Chem.,* 1976, **41**, 3505, 3512.

(b) Schmid, M. and Barner, R., *Helv. Chim. Acta,* 1979, **62**, 464.

(c) Fuganti, C. and Graselli, P., *J. Chem. Soc., Chem. Commun.,* 1979, 995.

Copper-Catalyzed Coupling of Grignard Reagents and Alkyl p-Tosylate Esters

(a) Fouquet, G. and Schlosser, M., *Angew. Chem., Int. Ed. Engl.,* 1974, **13**, 82.

(b) Schlosser, M., *Angew. Chem., Int. Ed. Engl.,* 1974, **13**, 701.

(c) Tamura, M. and Kochi, J., *Synthesis,* 1971, 303.

(d) Morisaki, M., Shibata, M., Duque, C., Imamura, N., and Ikekawa, N., *Chem. Pharm. Bull.,* 1980, **28**, 606.

(e) Schmidt, U., Talbiersky, J., Bartkowiak, F., and Wild, J., *Angew. Chem., Int. Ed. Engl.,* 1980, **19**, 198.

Synthesis 21:   cis-7,8-Epoxy-2-methyloctadecane (Disparlure),
A Gypsy Moth Sex Attractant

In the late 1860s the gypsy moth was imported into the U.S. from Europe and was inadvertently released. Larvae descending from those first moths have caused severe damage to trees in the Northeast by stripping leaves and thus killing entire forests. The female gypsy moth produces a pheromone (cf. syntheses 12 and 18) that attracts male gypsy moths and therefore facilitates reproduction. Chemical and field test experiments have shown that the major gypsy moth sex attractant is the dextrorotatory enantiomer *cis*-(7*R*,8*S*)-epoxy-2-methyloctadecane [(+)-disparlure, **21-1**]; (−)-disparlure is much less attractive to male gypsy moths. A gram-scale total synthesis of (+)-disparlure in greater than 98% enantiomeric purity has recently been reported, using as a key step organocopperlithium coupling with a chiral *p*-tosylate ester, as shown in eq. 21-1. Note that this organocopper substitution reaction proceeds chemospecifically without disturbing either the ether or the ester functional groups. How the key intermediate **21-2** was used in the synthesis of (+)-disparlure can be deduced from retrosynthetic scheme 21-1.

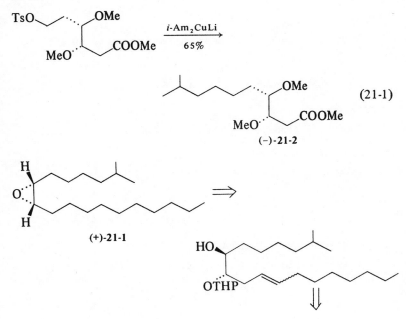

(21-1)

(−)-**21-2**

(+)-**21-1**

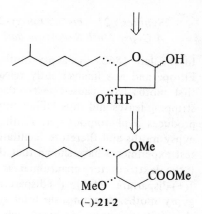

Retrosynthetic Scheme 21-1

References to Stoichiometric Organocopper Coupling with Alkyl Sulfonate Esters Since 1975

(+)-Disparlure

　　Mori, K., Takigawa, T., and Matsui, M., *Tetrahedron Lett.,* **1976,** 3953.

Prostaglandins

　　Stork, G. and Raucher, S., *J. Am. Chem. Soc.,* 1976, **98,** 1583.

trans,trans-Farnesol

　　Posner, G. H., Ting, J-S., and Lentz, C. M., *Tetrahedron,* 1976, **32,** 2281.

Optically Active 4-Alkyl-γ-lactones

　　Ravid, U. and Silverstein, M., *Tetrahedron Lett.,* **1977,** 423.

Acyclic Chiral Synthons

　　Trost, B. M. and Klun, T. P., *J. Am. Chem. Soc.,* 1979, **101,** 6756.

Optically Active α-Alkylcarboxylic Acids

　　Terashima, S., Tseng, C. C., and Koga, K., *Chem. Pharm. Bull.,* 1979, **27,** 747.

H. Allylic Acetates

Synthesis 22: (E,E)-8,10-Dodecadien-1-ol (Codlemone), A Codling Moth Sex Attractant

Codling moth damage of apple trees and other deciduous fruit trees is a worldwide problem. In the state of Washington alone, about 3 million dollars were spent annually during the 1970s on insecticides to prevent the codling moth from damaging apple orchards. The major codling moth sex attractant pheromone is (E,E)-8,10-dodecadien-1-ol (22-1).

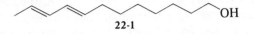

22-1

Retrosynthetic analysis of (E,E)-8,10-dodecadienol leads to a large number of reasonable synthetic approaches to this insect sex pheromone, and this long chain diene alcohol is a challenging target on which to refine your skill using the retrosynthetic process. The several syntheses of this pheromone published as of 1977 are summarized in an excellent review by Henrick (Tetrahedron, 1977, 33, 1). We focus attention in this chapter on preparation of (E,E)-8,10-dodecadienol using only organocopper coupling with allylic electrophiles.

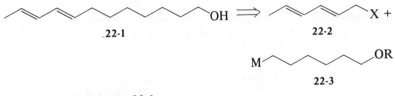

Retrosynthetic Scheme 22-1

The choice of leaving group X in allylic sorbyl-X species 22-2 is important (X = halide, alcohol, ether, carboxylate, or sulfonate), as is the choice of the metal (M) in organometallic species 22-3 [M = MgCl, MgBr, Cu(LiX), Cu(MgX$_2$), Cu · BF$_3$, or RuCuLi]. Serious potential problems during this coupling reaction involve loss of double bond geometry in the diene electrophile and allylic rearrangement, leading to branched-chain rather than straight-chain products. A full discussion of this reaction is given in Henrick's review. The general conclusions from an in-depth study of this reaction are as follows:

(1) organolithium (RLi) and organomagnesium (RMgX) reaction with sorbyl halides or sorbyl carboxylates leads to many (unidentified) products; (2) Li_2CuCl_4 -catalyzed reaction of a six-carbon *organolithium* reagent with six-carbon sorbyl acetate gave mainly sorbyl alcohol via organometallic attack at the acetate carbonyl group; (3) Li_2CuCl_4 -catalyzed reaction of a six-carbon *Grignard* reagent with sorbyl acetate gave a mixture of 1,2-dodecanediol (resulting from dimerization of the organometallic reagent) and (*E,E*)-8,10-dodeca-dienol which, after purification, was isolated in 40% yield (cf. synthesis 20); and (4) R_2CuLi (derived from a six-carbon RLi reagent) coupling with sorbyl acetate gave isomerically pure (*E,E*)-8,10-dodecadienol in 65% yield. Clearly, the diorganocopperlithium coupling is the cleanest, highest yield reaction. For relatively large-scale production of (*E,E*)-8,10-dodecadienol, however, the greater ease of Grignard rather than organolithium formation and the higher operating temperatures possible with the copper-catalyzed Grignard reaction rather than with the diorganocuprate coupling reaction combine to make the copper-catalyzed Grignard coupling more practical for industrial preparation of "codlemone" (22-1).

Organocopper Syntheses of Codling Moth Sex Attractant 22-1

Henrick, C. A., *Tetrahedron,* 1977, **33**, and references therein.
Samain, D. and Descoins, C., *Synthesis,* 1978, 388.

Descoins, C. and Henrick, C. A., *Tetrahedron Lett.,* 1972, 2999. *(Copper-Catalyzed Grignard Coupling with a Primary Alkyl Bromide)*

Mori, K., *Tetrahedron,* 1974, **30**, 3807. *(Copper-Catalyzed Grignard Coupling with a Primary Allylic Bromide)*

Copper-Catalyzed Grignard Coupling with Allylic Ethers

Claesson, A. and Olsson, L.-I., *J. Chem. Soc., Chem. Commun.,* 1978, 621.
Gendreau, Y. and Normant, J. F., *Tetrahedron,* 1979, **35**, 1517, and refences therein.
See also Yamaguchi, M., Murakami, M., and Mukaiyama, T., *Chem. Lett.,* 1979, 947.

Organocopper-Boron Complexes for Coupling with Allylic Alcohols and Allylic Carboxylates

> Yamamoto, Y., Yamamoto, S., Yatagai, H., and Maruyama, K., *J. Am. Chem. Soc.*, 1980, **102**, 2318. See also Ibuka, T. and Minakata, H., *Synth. Commun.*, 1980, **10**, 119.

Synthesis 23:   Periplanone-B, An American Cockroach Sex Attractant

In 1976 Dutch scientists isolated a total of 200 *micrograms* of periplanone-B (**23-1**) from 75,000 virgin female cockroaches! Periplanone-B is an intensely potent sex pheromone that functions mainly as a close proximity sex excitant for male cockroaches; threshold amounts for typical periplanone-B activity are 10^{-7} μg. In 1979 an American chemist synthesized periplanone-B for the first time; an important step in the total synthesis involved dimethylcopperlithium coupling with allylic acetate **23-2** (eq. 23-1). This coupling step involved substitution with allylic rearrangement, presumably because the more accessible end of the allylic acetate functionality was that one remote from the highly substituted cyclohexene ring. Retrosynthetic scheme 23-1 allows us to see in a general way how tetrasubstituted cyclohexene **23-3** was used in this creative approach to periplanone-B.

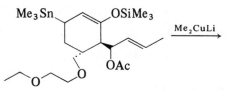

23-2

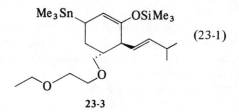

(23-1)

23-3

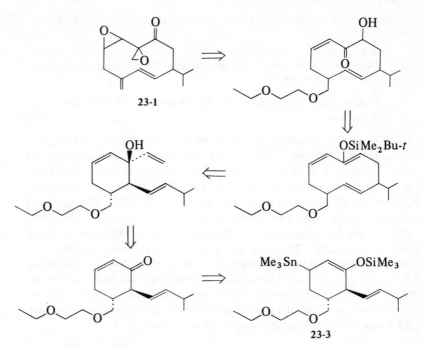

Retrosynthetic Scheme 23-1

Total Synthesis of (+)-Periplanone-B

Still, W. C., *J. Am. Chem. Soc.*, 1979, **101**, 2493.

Synthesis 24:   Insect Juvenile Hormones

We have discussed preparation of insect juvenile hormones via organocopper addition to acetylenes (synthesis 10) and via organocopper
coupling with alkenyl halides (synthesis 13). Here we note that organocopper reagents have also been used to prepare a Cecropia Moth juvenile hormone via coupling with allylic acetates. In a thorough study by
chemists at the Zoëcon Corporation in California, optimal reaction
conditions and carboxylate ester leaving groups were determined for
reaction 24-1, in which the new carbon-carbon double bond is formed

Table 24-1   Me$_2$CuLi Coupling with Esters 24-1 According to Eq. 24-1

Esters 24-1, OR =	Product (Relative Amounts)		
	10,11-trans-(24-2)	10,11-cis-	Direct Displacement
Trifluoroacetate	61	27	12
Acetate	93	5	2
Trimethylacetate	97	2	1
2,4,6-Trimethylbenzoate	97	2	1

predominantly with a *trans*-geometry; some of their results are summarized in Table 24-1.

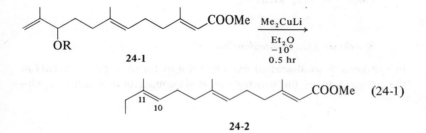

24-1

24-2

(24-1)

Having maximized the *trans*-stereochemistry of this substitution reaction with allylic rearrangement, the Zoëcon chemists hoped to use this procedure to introduce the central *trans* carbon-carbon double bond in Cepropia juvenile hormone **24-4** via coupling with allylic acetate **24-3** (eq. 24-2); although this reaction did proceed to give a nearly quantitative yield of allylically rearranged products, juvenile hormone **24-4** was formed in a disappointing 3:7 ratio along with its 6,7-*cis* double bond isomer. Note that under the reaction conditions of eq. 24-2 neither the β,β-disubstituted acrylate ester portion of the molecule nor the trisubstituted epoxide portion is disturbed; this chemospecificity of organocopper reagents is one of their most useful characteristics.

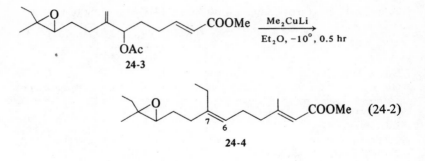

24-3

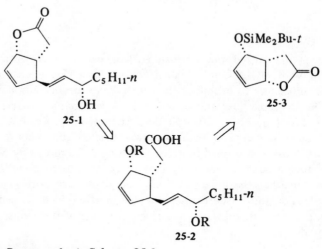

24-4

Zoëcon Synthesis of Juvenile Hormone 24-4:

Anderson, R. J., Henrick, C. A., Siddall, J. B., and Zurflüh, R., *J. Am. Chem. Soc.*, 1972, **94**, 5379.

Synthesis 25:   Prostaglandins

In synthesis 5 we discussed extensive use of organocopper reagents in the preparation of the prostaglandins via conjugate addition to cyclo-

25-1

25-3

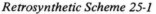

25-2

Retrosynthetic Scheme 25-1

pentenones. Here we focus on preparation of pivotal (i.e., convertible into the major prostaglandin types) prostaglandin intermediate **25-1** via organocopperlithium direct (rather than S_N2') displacement of an allylic carboxylate. Specifically, critical prostaglandin intermediate **25-1** can be analyzed in terms of protected hydroxy-acid **25-2** and then protected hydroxy-lactone **25-3** (retrosynthetic scheme 25-1.)

Although *ketal*-lactone **25-4** reacts with an R_2CuLi reagent to give comparable amounts of S_N2 and S_N2' products, *t*-butyldimethyl-silyl ether-lactone **25-3** reacts as in eq. 25-1 exclusively via S_N2-type coupling and stereospecifically with inversion of configuration at the carbon center undergoing substitution. Apparently, the bulky *t*-butyl-dimethylsilyl group sterically prevents the organocopper reagent from entering into an S_N2' (i.e., allylic transposition) displacement.

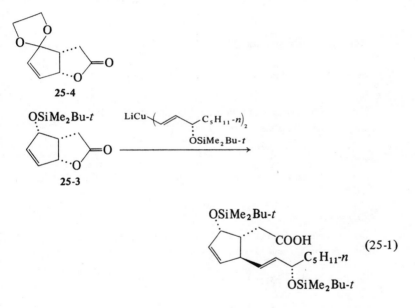

Summarizing organocopper reagent coupling with allylic acetates (syntheses 22–25), we note the following two generalizations: (1) a strong preference for attack at primary rather than secondary centers in unsymmetrical primary-secondary allylic acetates (e.g., **22-2** and **24-3**), and (2) a strong preference for attack at sterically more accessible secondary centers of secondary-secondary allylic acetates (e.g., **23-2** and **25-3**).

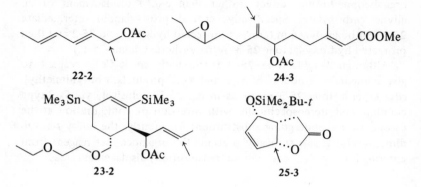

Retrosynthetic Scheme 25-1

Corey, E. J. and Mann, J., *J. Am. Chem. Soc.*, 1973, **95**, 6832.

Mechanistic Study of Organocopper Coupling with Allylic Oxygen Electrophiles

Allylic Acetates

Goering, H. L. and Singleton, V. D., Jr., *J. Am. Chem. Soc.*, 1976, **98**, 7854.

Allylic Ethers

(a) Gendreau, Y. and Normant, J. F., *Tetrahedron*, 1979, **35**, 1517.
(b) Claesson, A. and Olsson, L., *J. Chem. Soc., Chem. Commun.*, **1978**, 621.
(c) Calo, V., et al., *Tetrahedron Lett.*, **1979**, 3873.

References Since 1974 to Organocopper Coupling with Allylic Electrophiles

Allylic Acetates

(a) Mychajlowskij, W. and Chan, T., *Tetrahedron Lett.*, 1976, 4439.

(b) Cassani, G., Massardo, P., and Pricardi, P., *Tetrahedron Lett.*, **1979,** 633.

(c) Hammond, A. and Descoins, C., *Bull. Soc. Chim. Fr.*, **1978,** II-299.

(d) Trost, B. M. and Tanigawa, Y., *J. Am. Chem. Soc.*, 1979, **101,** 4413.

Allylic Epoxides

(a) Teutsch, G. and Bélanges, A., *Tetrahedron Lett.*, **1979,** 2051.

(b) Marino, J. P. and Floyd, D. M., *Tetrahedron Lett.*, **1979,** 675.

(c) Marino, J. P. and Hatanaka, N., *J. Org. Chem.*, 1979, **44,** 4467.

(d) Chapleo, C. B., Finch, M. A. W., Lee, T. V., Roberts, S. M., and Newton, R. F., *J. Chem. Soc., Chem. Commun.*, **1979,** 676.

Allylic Alcohols

(a) Tanigawa, Y., Kanamaru, H., Sonoda, A., and Murahashi, S., *J. Am. Chem. Soc.*, 1977, **99,** 2361.

(b) Yamamoto, Y. and Maruyama, K., *J. Organomet. Chem.*, 1978, **156,** C9.

(c) Yamamoto, Y., Yamamoto, S., Yatagai, H., and Maruyama, K., *J. Am. Chem. Soc.,* 1980, **102,** 2318.

Allylic Ethers

(a) Gendreau, Y. and Normant, J. F., *Bull. Soc. Chim. Fr.*, **1979,** II-305.

(b) Normant, J. F., Commercon, A., Gendreau, Y., Bourgain, M., and Villieras, J., *Bull. Soc. Chim. Fr.*, **1979,** II-309.

(c) Calo, V., Lopez, L., Marchese, G., and Pesce, G., *Synthesis*, **1979,** 885.

(d) Yamaguchi, M., Murakami, M., and Mukaiyama, T., *Chem. Lett.,* **1979,** 957.

Allylic Chlorides

Mura, A. J., Jr., Bennett, D. A., and Cohen, T., *Tetrahedron Lett.*, **1975,** 4433.

Allylic Halides

(a) Seebach, D. and Lehr, F., *Helv. Chim. Acta*, 1979, **62,** 2239.

(b) Chenard, B. L., Manning, M. J., Raynolds, P. W., and Swenton, J. S., *J. Org. Chem.*, 1980, **45,** 378.

Allylic Selenides

Reich, H. J., *J. Org. Chem.,* 1975, **40,** 2570.

Allylic Sulfoxides and Sulfones

Masaki, Y., Sakuma, K., and Kaji, K., *Chem. Pharm. Bull.*, 1980, 434.

Allylic Ammonium Iodides

Dressaire, G. and Langlois, Y., *Tetrahedron Lett.*, 1980, **21**, 67; see also Posner, G. H. and Ting J-S., *Synth. Commun.*, 1974, **4**, 335.

I. Propargylic Acetates

Synthesis 26: An Allenic Prostaglandin

Because of their extraordinarily high and general biological activity, natural prostaglandins have stimulated an enormous effort by organic chemists to synthesize prostaglandin structural analogs, hoping and expecting that one or more of these synthetic analogs could be developed into a useful, specific new drug (cf. syntheses 5 and 25).

Prostaglandin allene **26-3**, derived directly from allene **26-2**, is a biologically active (luteolytic) analog prepared via organocopper reaction with propargylic acetate **26-1** via a 1,3-substitution path (eq. 26-1).

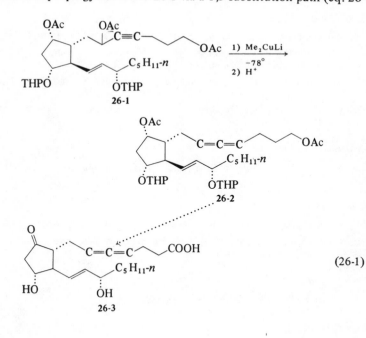

$$(26\text{-}1)$$

Mechanistically, the 1,3-substitution process is thought to involve an organocopper nucleophilic displacement with allylic transposition of the propargylic electrophile and formation of a reactive allenic copper (III) intermediate (eq. 26-2). The intermediate has been trapped at low temperature by reaction with various electrophilic species (including H_2O, D_2O, I_2, and carbon compounds); at higher temperature it undergoes a reductive elimination with formation of a new carbon-carbon bond. Although the most common use of this substitution process is to form new carbon-carbon bonds, eq. 26-1 represents one example in which hydrolysis of the allenic copper intermediate leads to a desired *un*alkylated allenic prostaglandin.

In some cases when the propargylic acetate contains a bulky group Ra (e.g., Ra = Me$_3$Si, in eq. 26-2), then organocopper substitution can occur *directly* at the propargylic carbon atom to give an acetylenic rather than an allenic product.

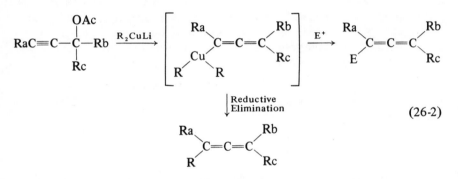

(26-2)

Synthesis of Allenic Prostaglandin 26-2

Crabbé, P. and Carpio, H., *Chem. Commun.*, 1972, 904.

Synthesis of 1-Hydroxyvitamin D via Organocopper-Propargylic Acetate Coupling

Hammond, M. L., Mourino, A., and Okamura, W. H., *J. Am. Chem. Soc.*, 1978, **100**, 4907.

Recent References to Organocopper Coupling with Propargylic Acetates

Luche, J. L., Barreiro, E., Dollat, J. M., and Crabbé, P., *Tetrahedron Lett.*, 1975, 4615.

Crabbé, P., Barreiro, E., Dollat, J. M., and Luche, J.-L., *J. Chem. Soc., Chem. Commun.*, 1976, 183.

Dollat, J. M., Luche, J.-L., and Crabbé, P., *J. Chem. Soc., Chem. Commun.*, 1977, 761.

Brinkmeyer, R. S. and MacDonald, T. L., *J. Chem. Soc., Chem. Commun.*, 1978, 876.

Synthesis 27:   (-)-Methyl (E)-2,4,5-tetradecatrienoate, A Male Bean Weevil Sex Attractant

The male bean weevil sex attractant **27-1** [(−)-methyl (*E*)-2,4,5-tetra-decatrienoate] is a pheromone (see syntheses 12, 18, 21, 22, and 23) produced by female bean weevils. Because of the extensive agricultural damage done by bean weevils, control of these insect pests is an important task, and bean weevil pheromones are beginning to play a significant part in this process.

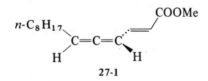

27-1

As an optically active allene, natural pheromone **27-1** has been synthesized with good optical purity using an organocopper reagent coupling with propargylic alcohol derivative **27-2** (eq. 27-1). Just as organocopper reagents react in a highly stereocontrolled fashion with optically active secondary halides and tosylates to form products with inverted configuration, they also react with various optically active propargylic alcohol derivatives (OAc, OCONHR, OSOAr) to give chiral allenes via either a *syn* or *anti* 1,3-substitution process. The *syn* 1,3-substitution reaction has been noted for those cases in which the chiral propargylic carbon atom is part of a carbocycle (e.g., 17-ethynyl-17-hydroxy steroids), and the *anti* 1,3-substitution pathway has been found in acyclic propargylic systems. Reaction 27-1 was found to proceed under optimal conditions to give the chiral (R)-allene in excellent chemical yield and in 77% enantiomeric excess [88.5% (R) and 11.5% (S) enantiomers]. The degree of stereocontrol in these displacement reactions often depends in a sensitive way on the order of reagent mixing, the nature of the leaving group, and the solvent.

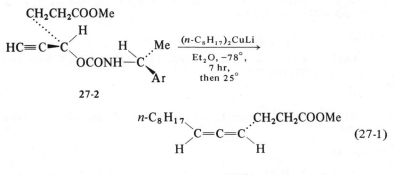

27-2

$$(27\text{-}1)$$

27-3

Subsequent conversion of ester **27-3** into α,β-ethylenic ester **27-1** was achieved smoothly via α-phenylselenation, oxidation to the selenoxide, and pyrolytic elimination of PhSeOH.

Synthesis of Pheromone 27-1

Pirkle, W. H. and Boeder, C. W., *J. Org. Chem.*, 1978, **43**, 2091.

Descoins, C., Henrick, C. A., and Siddall, J. B., *Tetrahedron Lett.*, **1972**, 3777.

Organocopper Coupling with Propargylic-X Species

X = Sulfonate

Doutheau, A., Balme, G., Malacria, M., and Goré, J., *Tetrahedron Lett.*, **1978**, 1803.

Verkruijese, H. D. and Hasselaar, M., *Synthesis*, **1979**, 292

Montury, M., Psaume, B., and Goré, J., *Tetrahedron Lett.*, 1980, **21**, 163.

X = Sulfinate

Westmijze, H. and Vermeer, P., *Synthesis*, 1979, 390.

Kleijn, H., Westmijze, H., Kruithof, K., and Vermeer, P., *Rec. Trav. Chim. Pays-Bas*, 1979, **38**, 27.

Westmijze, H. and Vermeer, P., *Tetrahedron Lett.*, 1979, 4101.

Klein, H., Elsevier, C. J., Westmijze, A., Meijer, J., and Vermeer, P., *Tetrahedron Lett.*, 1979, 3101.

X = OMe

Claesson, A. and Sahlberg, C., *Tetrahedron Lett.*, 1978, 1319; *J. Organomet. Chem.*, 1979, **170**, 355.

X = OTHP

Claesson, A., Tammeforg, I., and Olsson, L. I., *Tetrahedron Lett.*, 1975, 1509.

X = Epoxide

Vermeer, P., Meijer, J., de Graaf, C., and Schrewis, H., *Rec. Trav. Chim. Pays-Bas*, 1974, **93**, 46.

X = Halide

Näf, F., Decorzant, R., Thommen, W., Willhalm, B., and Ohloff, G., *Helv. Chim. Acta*, 1975, **58**, 1016.

Eiter, K., Lieb, F., Disselnkötter, H., and Oediger, H., *Ann.*, 1978, 658.

Pasto, D. J., Chou, S. K., Waterhouse, A., Shults, R. H., and Hennion, G. F., *J. Org. Chem.*, 1978, **43**, 1385, 1389.

Luteijn, J. M. and Spronck, H. J. W., *J. Chem. Soc., Perkin I*, 1979, 201.

Savignac, P., Breque, A., Charrier, C., and Matthey, F., *Synthesis*, 1979, 832.

FeCl$_3$ Catalysis

Pasto, D. J., Hennion, G. F., Shults, R. H., Waterhouse, A., and Chou, S. K., *J. Org. Chem.*, 1976, **41**, 3496.

J. Epoxides

Synthesis 28: *erythro-3,7-Dimethylpentadec-2-yl Acetate and Propionate, Pine Sawfly Sex Pheromones*

Pine trees are usually resistant to insect attack because most insects are repelled by pine oil and resin. Pine sawfly larvae, however, ingest large quantities of pine needles, separating and retaining the pine oil and resin, and subsequently using these repellents against their own enemies! Protection of pine forests from pine sawfly infestation and damage therefore requires some ecologically acceptable chemical means for controlling pine sawfly populations; as discussed in syntheses 12, 18, 21, 22, 23, and 27, many species-specific sex attractants (phero-mones) are being synthesized and field-tested for effective insect pest control management.

erythro-3,7-Dimethylpentadecan-2-ol (**28-1**), in the form of its acetate and its propionate, is found to be the sex attractant of two genera of pine sawflies. Although there are four possible stereoisomers of *erythro* alcohol **28-1**, it has been suggested that the major biological activity of the natural sex attractant may reside in only one stereoisomer. The first step in testing this suggestion is to prepare all four possible stereoisomers in optically pure form. This synthetic task has recently been achieved, but as of this writing the results of the biological testing of each pure stereoisomer are not yet available.

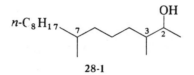

28-1

The crucial step in the stereocontrolled syntheses reported for *erythro* alcohols **28-1** involves nucleophilic *trans*-opening of both enantiomers of optically active *trans*-2-butene oxide by both enan-tiomers of an organocopperlithium reagent (eqs. 28-1 and 28-2). The degree of stereocontrol in this organocopper displacement reaction (C-O $\longrightarrow$ C-C) approaches 100%. Therefore the advantages of organo-copper reagents over organomagnesium and organolithium reagents for epoxide opening with concomitant carbon-carbon bond formation, first publicized by Johnson and co-workers in 1970, include dramatic and reproducible stereospecificity.

The optically active organocopper reagents and the optically active epoxides in eqs. 28-1 and 28-2 were prepared from natural chiral citronellol and natural chiral tartaric acid, respectively.

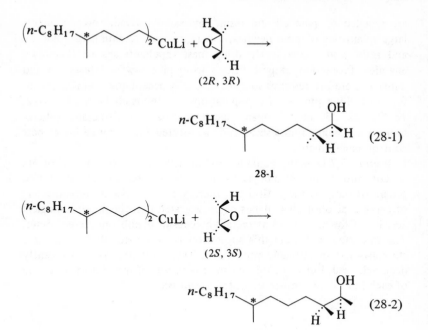

$$(2R, 3R)$$

$$(28\text{-}1)$$

28-1

$$(2S, 3S)$$

$$(28\text{-}2)$$

Total Synthesis of Optically Pure Enantiomers of *Erythro* Alcohol 28-1

Mori, K., Tamada, S., and Matsui, M., *Tetrahedron Lett.,* **1978**, 901.

Mori, K. and Tamada, S., *Tetrahedron,* **1979**, **35**, 1279.

See also Mori, K., and Iwasawa, H., *Tetrahedron,* **1980**, **36**, 87.

See also Baker, R., Winton, P. M., and Turner, R. W., *Tetrahedron Lett.,* **1980**, **21**, 1175.

Organocopper Opening of Epoxides

Posner, G. H., Ph.D. thesis, Harvard University, **1968**, 89. *Diss. Abstr.,* **1968**, **29**, 1613-B.

Herr, R. W., Wieland, D. M., and Johnson, C. R., *J. Am. Chem. Soc.,* **1970**, **92**, 3813.

Anderson, R. J., *J. Am. Chem. Soc.,* **1970**, **92**, 4978.

Staroscik, J. and Rickborn, B., *J. Am. Chem. Soc.,* **1971**, **93**, 3046.

Hudrlik, P. F., Peterson, D., and Rona, R. J., *J. Org. Chem.*, 1975, **40**, 2263.

Elliot, W. J. and Fried, J., *J. Org. Chem.*, 1976, **41**, 2469.

Acker, R. D., *Tetrahedron Lett.*, **1977**, 3407.

Johnson, M. R., Nakata, T., and Kishi, Y., *Tetrahedron Lett.*, 1979, 4343.

Anderson, R. J., Adams, K. G., Chinn, H. R., and Henrick, C. A., *J. Org. Chem.*, 1980, **45**, 2229.

Copper-Catalyzed Grignard Openings

Linstrumelle, G., Lorne, R., and Dang, H. P., *Tetrahedron Lett.*, **1978**, 4069.

Huynh, C., Derguini-Goumechal, F., and Linstrumelle, G., *Tetrahedron Lett.*, **1979**, 1503.

Kelly, A. G. and Roberts, J. S., *J. Chem. Soc., Chem. Commun.*, **1980**, 228.

Atkins, M. P., Golding, B. T., Bury, A., Johnson, M. D., and Sellars, P. J., *J. Am. Chem. Soc.*, 1980, **102**, 3630.

Vinylic Epoxides

Teutsch, G. and Bélanges, A., *Tetrahedron Lett.*, 1979, 2051.

Marino, J. P., and Hatanaka, N., *J. Org. Chem.*, 1979, **44**, 4467.

Cahiez, G., Alexakis, A., and Normant, J. F., *Synthesis*, 1978, 528.

Aziridines

Kozikowski, A. P., Ishida, H., and Isobe, K., *J. Org. Chem.*, 1979, **44**, 2788.

Synthesis 29:   A 2-C-Methyl Pyranoside, A Sugar

Although natural sugars (carbohydrates) are essential to human health, structurally modified synthetic sugars are being prepared and used more and more often as optically active building blocks for laboratory construction of such complex chiral organic compounds as prostaglandins and macrolide antibiotics. Because of the very strong tendency of organocopper reagents to open epoxides with inversion of configuration (i.e., backside attack) and because of the well established preference for *trans*-diaxial opening of cyclohexene type oxides, 2,3-epoxy pyranoside **29-1** was expected and indeed was found to react cleanly with dimethylcopperlithium to give methylated pyranoside **29-2** in good yield (eq. 29-1). The nearly exclusive formation of alcohol **29-2** stands in sharp contrast to most other ring-scission reactions of carbohydrate epoxides, in which many different products are usually produced. Thus eq. 29-1 exemplifies the high chemoselectivity of organocopperlithium reagents and their great utility in coupling reac-

tions even when the electrophile contains many (basic) oxygen atoms
and reacts easily via various unwanted reaction pathways.

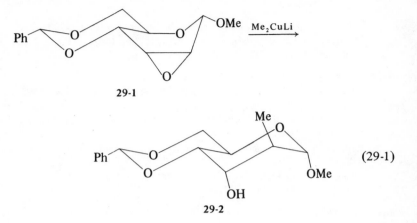

29-1

(29-1)

29-2

Eq. 29-1

Hicks, D. R., Ambrose, R., and Fraser-Reid, B., *Tetrahedron Lett.,* **1973,**
2507.
Hanessian, S. and Rancourt, G., *Can. J. Chem.,* 1977, **55,** 1091.

Alkylcitric Acids

Brandiange, S., *Acta Chem. Scand.,* 1977, **B31,** 307.

(-)-N-Methylmaysenine

Corey, E. J., Weigel, L. O., Chamberlin, A. R., and Lipshutz, B., *J. Am.
Chem. Soc.,* 1980, **102,** 1439.

*Synthesis 30: Side Chain of Deoxyharringtonine, An
Antileukemia Alkaloid*

We have discussed alkaloids in synthesis 7 and cancer chemotherapy
in synthesis 6. One of the natural plant extracts that is very active

against experimental lymphoid leukemia in mice is harringtonine (**30-1**). Currently even more promising in terms of cancer chemotherapy is homoharringtonine (**30-2**), in which the diacid side chain has three contiguous methylene groups instead of the natural two; homoharringtonine is one of the synthetic compounds in development at the National Cancer Institute of the U.S. National Institutes of Health as a potential drug for cancer chemotherapy. If the side-chain diacid group is removed from harringtonine, the alkaloid cephalotaxine (**30-3**) is formed. It is noteworthy that neither cephalotaxine itself nor the diacid side chain itself shows anticancer activity. One of the most direct and, in principle, simple synthetic approaches to harringtonine is to esterify cephalotaxine with the appropriate diacid side chain. Toward this end several laboratory approaches have been developed for preparation of cephalotaxine and of the diacid side chain. We focus attention here on one route to diester **30-4**, characteristic of the deoxy side chain of harringtonine, involving organocopper nucleophilic opening of an unsymmetrically substituted epoxide.

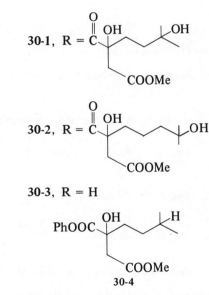

$$30\text{-}1, \ R = \overset{\overset{\text{O}}{\|}}{\text{C}}$$

$$30\text{-}2, \ R = \overset{\overset{\text{O}}{\|}}{\text{C}}$$

30-3, R = H

30-4

Unsymmetrical epoxide **30-5** reacts with diisobutylcopperlithium exclusively at the unsubstituted epoxide carbon atom to form tertiary alcohol **30-4**. Clearly, this route also allows easy access to the side chain of deoxyhomoharringtonine simply by using the homologous organo-copper reagent, diiso*amyl*copperlithium. The functional group selectivity (chemoselectivity) of this displacement reaction is typical of organo-copper chemistry; neither the benzyl ester nor the methyl ester group is disturbed (eq. 30-1).

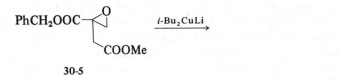

30-5

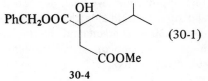

(30-1)

30-4

Synthesis of Side chain 30-4

Auerbach, J., Ipaktchi, T., and Weinreb, S., *Tetrahedron Lett.,* 1973, 4561.

Review on Synthesis of Cephalotaxine (30-3)

Weinreb, S. M. and Semmelhack, M. F., *Acc. Chem. Res.,* 1975, 8, 158.

Synthesis of Lasalocid A

Ireland, R. E., Thaisrivongs, S., and Wilcox, C. S., *J. Am. Chem. Soc.,* 1980, 102, 1157.

Synthesis 31: Insect Juvenile Hormones

We have discussed insect juvenile hormones in syntheses 10, 13, and 24. Because of the relatively easy preparation of bisepoxide **31-1** from readily available (*E,E*)-farnesol, Yamamoto and Sharpless designed a synthesis of 18-carbon *Cepropia* juvenile hormone **31-2** involving a key organocopper opening of both epoxide rings of bisepoxide **31-1** regiospecifically at the less substituted carbon centers (eq. 31-1). Subsequent transformation of the glycol groups into trisubstituted alkenes and finally terminal epoxidation led to 18-carbon juvenile hormone **31-2**.

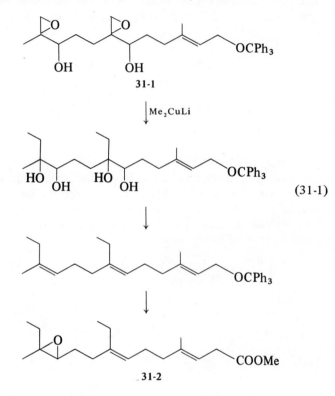

(31-1)

Synthesis of 18-Carbon Juvenile Hormone 31-2

Tanaka, S., Yamamoto, H., Nozaki, H., Sharpless, K. B., Michaelson, R. C., and Cutting, J. D., *J. Am. Chem. Soc.,* 1974, 96, 5254.

Yasuda, A., Tanaka, S., Yamamoto, H., and Nozaki, H., *Bull. Chem. Soc. Japan,* 1979, 52, 1701.

Synthesis 32: Prostaglandins

We have discussed prostaglandins in syntheses 5, 25, and 26, and we have noted in syntheses 5 and 25 that attachment of the omega chain (carbon-13 through carbon-20) via nucleophilic attack on an electrophilic cyclopentane ring can occur not only at the desired electrophilic carbon center but also occasionally at an undesired electrophilic position (cf. **5-9** and **25-3**).

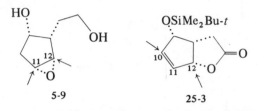

Likewise, epoxycyclopentane **32-1** can suffer nucleophilic attack at either carbon-11 or carbon-12 (prostaglandin numbering). It was discovered that divinylcopperlithium (but not most other nucleophilic reagents) reacts with epoxide **32-1** regioselectively at desired position 12 to produce the Corey lactol **32-2** and therefore to form ultimately various prostaglandins (eq. 32-1).

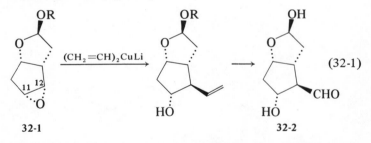

Understanding what factors direct nucleophilic attack of the organocopper reagent so selectively toward electrophilic position 12 rather than position 11 is difficult. A very similar approach, however, has been reported using epoxycyclopentane ketal **32-3**; a mixed (acetylenic)-(vinylic)cuprate caused selective opening of epoxide **32-3** at position 12 with introduction of the entire eight-carbon omega chain of the prostaglandins (eq. 32-2) rather than introduction of just a two-carbon vinyl groups as in eq. 32-1.

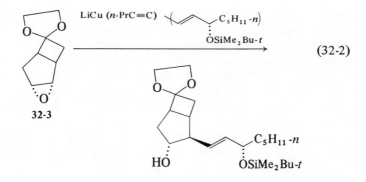

(32-2)

32-3

Eq. 32-1

Corey, E. J., Nicolaou, K. C., and Beames, D. J., *Tetrahedron Lett.*, 1974, 2439.

Eq. 32-2

Newton, R. F., Howard, C. C., Reynolds, D. P., Wadsworth, A. H., Crossland, N. M., and Roberts, S. M., *J. Chem. Soc., Chem. Commun.*, 1978, 662.

Cave, R. J., Howard, C. C., Klinkert, G., Newton, R. F., Reynolds, D. P., Wadsworth, A. H., and Roberts, S. M., *J. Chem. Soc., Perkin I*, 1979, 2954.

Howard, C. C., Newton, R. F., Reynolds, D. P., Wadsworth, A. H., Kelly, D. R., and Roberts, S. M., *J. Chem. Soc. Perkin I*, 1980, 852.

Organocopper Opening of an Epoxycyclopentane Leading to Muscarine Analogs

Givens, R. S., Rademacher, D. R., Kongs, J., and Dickerson, J., *Tetrahedron Lett.*, 1974, 3211.

K. Oxidative Dimerization

Synthesis 33: Kotanin, A Mold Metabolite

Most organocopper reagents are extremely sensitive to heat and to oxidants. Ethylcopper, for example, decomposes at $-18°$ and phenyl-

copper at 80°; thermal stability decreases as follows: neopentyl >
methyl > n-propyl > ethyl > isopropyl. Perhaloakylcopper and per-
haloarylcopper species are much more thermally stable than the corresponding nonhalogenated species. Complete oxidative decomposition
of diorganocopperlithium reagents can be achieved rapidly even at
-78° simply by bubbling oxygen into the dilute (≤0.1 M) reaction
mixture. Other oxidants [e.g., nitroaromatics and copper(II) salts]
have also been used effectively.

A large number of RCu and RCu · ligand species have been thermally or oxidatively dimerized to symmetrical products R-R. Examples
include dimerization of alkyl, alkenyl, alkynyl, aryl, heteroaryl,
benzylic, and functional alkyl groups. There are several outstanding
features of these dimerizations. First, thermolysis or oxidation of
alkenylcopper species produces dimers (butadiene derivatives) stereospecifically with retention of configuration. Second, dimerization of
functionalized alkylcopper reagents has led to preparation of various
unusual types of compounds (e.g., 1,2-bissulfones, 1,2-bisphosphine
oxides) and to fundamental classes of compounds having substantial
synthetic utility (e.g., 1,4-diketones, eq. 33-1). Finally, dimerization
of organocopper species may be the preferred method to prepare
natural products accessible only with difficulty by other means (e.g.,
mold metabolite kotanin, **33-1**, eq. 33-2).

$$(C_6H_5)_2C{=}NN \qquad\qquad (C_6H_5)_2C{=}NN \qquad NN{=}C(C_6H_5)_2$$
$$\underset{RCCH_2Cu}{\overset{\|}{}} \xrightarrow{0\text{-}20°} \underset{RCCH_2CH_2CR}{\overset{\|}{}}$$

$$\xrightarrow{H_2O} \quad R\overset{O}{\overset{\|}{C}}CH_2CH_2\overset{O}{\overset{\|}{C}}R \quad (33\text{-}1)$$

$$2,4,6\text{-}(CH_3O)_3C_6H_2Cu \xrightarrow[(25\%)]{CuCl_2} 2,4,6\text{-}(CH_3O)_3C_6H_2C_6H_2(OCH_3)_3\text{-}2,4,6$$

33-1

(33-2)

Relatively few lithium diorganocuprate(I) compounds have been
intentionally thermolyzed or oxidized. Several examples are available
of alkyl, alkenyl, alkynyl, and aryl group dimerizations. The most
useful feature of these dimerizations is that typical radicals are not

involved; thus dineophylcopperlithium (eq. 33-3) and dialkenylcopper-lithium reagents are thermally and oxidatively dimerized without structural or geometric rearrangement.

$$[C_6H_5C(CH_3)_2CH_3]_2CuLi \xrightarrow[O_2,-78°]{THF} C_6H_5C(CH_3)_2CH_2CH_2C(CH_3)_2C_6H_5$$

(33-3)

Reviews on Oxidative Dimerization of Organocopper Reagents

(a) Kauffmann, T., *Angew. Chem., Int. Ed. Engl.*, 1974, **13**, 291.

(b) Posner, G., *Org. React.*, 1975, **22**, 253.

References to Organocopper Dimerization Since 1974

Copper(I) Induced Dimerization

(a) Cohen, T. and Cristea, I., *J. Org. Chem.*, 1975, **40**, 3649.

Bromoallene Dimerization to Conjugated Diallenes

(b) Toda, F. and Takehira, Y., *J. Chem. Soc., Chem. Commun.*, **1975**, 174.

Methylcopper Induced Dimerization of Dialkenylchloroboranes

(c) Yamamoto, Y., Yatagai, H., and Moritani, I., *J. Am. Chem. Soc.*, 1975, **97**, 5606.

Oxidative Dimerization

(d) Banks, R. B. and Walborsky, H. M., *J. Am. Chem. Soc.*, 1976, **98**, 3732.

Cupric Chloride-Promoted Dimerization of 1-Cyclopropenylcopper Species

(e) Sorokin, V. I., *Russ. J. Org. Chem.*, 1977, **13**, 673.

Copper(0)-Isonitrile-Induced Dimerization of Alkyl Halides

(f) Ballatore, A., Crozet, M. P., and Surzur, J.-M., *Tetrahedron Lett.*, **1979**, 3073.

4

Laboratory
Procedures

Because so many different conditions have been used in successful organocopper reactions and because such detailed review articles (see p. 8) have been written on this subject, we include in this introductory book only a few experimental procedures to illustrate how simple most organocopper reactions are to set up and perform.

A. *Phenyl t-Butyl Ketone* [Use of a PhS(*t*-Bu)CuLi heterocuprate]

To a clean, dry 25-ml, three-necked, round-bottomed flask fitted with two serum caps and a nitrogen-filled balloon, and containing a magnetic stirring bar and phenylthiocopper(I) (0.242 g, 1.40 mmol, prepared according to Adams, R., Reifschneider, W., and Ferretti, A., *Org. Synth.*, 1962, **42**, 22, and also now commercially available from the Alpha Division of the Ventron Corp.), was added dry tetrahydrofuran (10 ml). After cooling to -20°, *t*-butyllithium (1.39 mmol) was added, forming a clear solution. After 5 min the solution was cooled to -78°, and benzoyl chloride (0.241 g, 1.00 mmol) was added. After stirring for 1 hr at -78°, the reaction was quenched with absolute methanol (1 ml) and was allowed to warm to room temperature. The reaction mixture was added to saturated aqueous ammonium chloride (50 ml) and was stirred for 15 min. After the phenylthiocopper was removed by suction filtration, the organic products were extracted into three portions of ether (3 × 25 ml). The ether layers were dried over magnesium sulfate, filtered, and rotary-evaporated to give the crude phenyl t-butyl ketone (pivalophenone); yield: 0.160 g (95%). Microdistillation afforded colorless pivalophenone having identical spectral and physical properties to those of an authentic sample; yield: 0.135 g (80%); bp 105–106°/15 torrs [Posner, G. H., Brunelle, D. J., and Sinoway, L., *Synthesis*, **1974**, 662; cf. Posner, G. H. and Whitten, C. E., *Org. Synth.*, 1976, **55**, 122].

B. *Preparation of Me₂S · CuBr Complex* [Avoiding side reactions from presence of Cu(II) species and other metal salt impurities]

To 40.0 g (279 mmol) of pulverized cuprous bromide (Fisher Scientific, or according to Keller, R. N., and Wycoff, H. D., *Inorg. Syntheses,* 1946, **2**, 1) was added 50 ml (42.4 g, 682 mmol) of dimethyl sulfide (Eastman, bp 36–38°). The resulting mixture, which warmed during dissolution, was stirred vigorously and then filtered through a glass wool plug. The residual solid was stirred with an additional 30 ml (25 g, 409 mmol) of dimethyl sulfide to dissolve the bulk of the remaining solid, and this mixture was filtered. The combined red solutions were diluted with 200 ml of hexane. The white crystals that separated were filtered with suction and washed with hexane until the washings were colorless. The residual solid was dried under nitrogen to leave 51.6 g (90%) of the complex as white prisms that dissolved in a diethyl ether-dimethyl sulfide mixture to give a colorless solution. For recrystallization a solution of 1.02 g of the complex in 5 ml of dimethyl sulfide was slowly diluted with 20 ml of hexane to give 0.96 g of the pure complex as colorless prisms, mp 124–129° dec. The complex is essentially insoluble in hexane, diethyl ether, acetone, chloroform, carbon tetrachloride, methanol, ethanol, and water. Although the complex does dissolve in dimethylformamide and in dimethyl sulfoxide the facts that heat is evolved and the resulting solutions are green suggest that the complex has dissociated and that some oxidation (or disproportionation) to give Cu(II) species has occurred. (House, H. O., Chu, C.-Y., Wilkins, J. M., and Umen, M. J., *J. Org. Chem.,* 1975, **40**, 1460).

C. *5-Methyl-1,4-hexadiene* [Use of Me₂S · CuBr in an organocopper addition to a terminal acetylene followed by reaction of the intermediate vinylcopper species with an added electrophile]

A mixture of dimethyl sulfide–cuprous bromide complex (6.15 g, 30 mmol), ether (35 mL), and dimethyl sulfide (30 mL) under nitrogen was cooled to −45°C, and a 2.95 M solution of methylmagnesium bromide (10.35 mL, 30 mmol) was added dropwise. After the resulting suspension of yellow solid was stirred at −45°C for 2 hr, propyne (2.0 mL, 35 mmol) which had been condensed at −50°C under nitrogen was added with a dry ice cooled syringe over a 1 min period. The reaction mixture was stirred at −23°C for 120 hr. After the resulting dark green solution was cooled to −78°C, HMPT (hexamethylphosphoric triamide, 10.4 mL, 60 mmol) and allyl bromide (3.0 mL, 34 mmol) were added separately. The mixture was then stirred at −30°C for 12 hr, warmed to 0°C, quenched with a saturated aqueous ammonium chloride solution (adjusted to pH 8 with ammonia), and partitioned between additional ether and water. The ether layer was washed with additional aqueous ammonium chloride, water, and saturated aqueous sodium chloride, dried over anhydrous magnesium sulfate, and partially concentrated in vacuo. The crude reaction product was distilled to give 2.0 g (70%) of 100% pure 5-methyl-1,4-hexadiene, bp 88–89°C (760 torrs) (Marfat, A., McGuirk, P. R., and Helquist, P., *J. Org. Chem.*, 1979, **44**, 3888).

D. *2,6-Dimethyl-2,6-heptadien-1-ol* [Copper catalysis]

Under a nitrogen atmosphere an ethereal solution of Me₂C=CHLi (30 mmol) is added at −25° to a mixture of

$$CH_2=CHC-CH_2 \quad \overset{\displaystyle CH_3}{\underset{\displaystyle O}{\overset{|}{\diagdown\diagup}}}$$

(25 mmol) and cuprous bromide (1.25 mmol) in 100 ml of diethyl ether. After stirring for 1.5 hr at −15°, the reaction mixture is hy-

drolyzed with 50 ml of saturated aqueous ammonium chloride. The aqueous layer is extracted twice with ether (250 ml). The extracts and the organic layer are washed with water (50 ml) and dried with magnesium sulfate. Distillation gives pure 2,6-dimethyl-2,6-heptadien-1-ol in 94% yield as an 88:12 $E:Z$ mixture of geometric isomers, bp 105° (15 torrs). [Cahiez, C., Alexakis, A., and Normant, J. F., *Synthesis*, **1978**, 528].

Author Index

Subject Index

Index of Retrosynthetic Schemes

Compound Index